DESTINATION MOON

A History of the Lunar Orbiter Program

CONTENTS

		Page
Preface		
I	Unmanned Lunar Exploration and the Need for a Lunar Orbiter	1
II	Toward a Lightweight Lunar Orbiter	9
III	Beginning the Lunar Orbiter Program	49
IV	NASA and Boeing Negotiate a Contract	75
V	Implementing the Program	97
VI	The Lunar Orbiter Spacecraft	111
VII	Building the Spacecraft: Problems and Resolutions	133
VIII	Lunar Orbiter Mission Objectives and Apollo Requirements	177
IX	Missions I, II, and III: Apollo Site Search and Verification	225
X	Missions IV and V: The Lunar Surface Explored	269
XI	Conclusions: Lunar Orbiter's Contribution to Space Exploration	303
XII	Lunar Orbiter Photography	331
Appendix A	Glossary	361
Appendix B	Organization Charts	363
Appendix C	Record of Unmanned Lunar Probes, 1958-1968	367
References		375

PREFACE

In June 1967, as a member of the NASA History Office Summer Seminar, I began work on a history of the Lunar Orbiter Program, then in its operational phase. My objective was to document the origins of the program and to record the activity of the missions in progress. I also wanted to study the technical and management aspects of the lunar orbital reconnaissance that would provide the Apollo Program with photographic and selenodetic data for evaluating the proposed astronaut landing sites.

Lunar Orbiter brought several new departures in U.S. efforts to explore the Moon before landing men there. It was the first big deep space project for Langley Research Center. It came into being in 1963 after the Ranger and Surveyor Programs were well along in their development and at a time when the data it could acquire would be timely to Apollo only for mission design, not for equipment design, since the decisions on the basic Apollo equipment had already been made. Although Lunar Orbiter was not a "crash" effort, it did require that Langley Research Center set up a development and testing schedule in which various phases of the project would run nearly concurrently. This approach had not been tried before on a major lunar program.

Research led me first to the Office of Space Science and Applications at NASA Headquarters in Washington. I discussed the project with Lunar Orbiter Program officials and received help and encouragement from Oran W. Nicks, the Director of Lunar and Planetary Programs (later Deputy Director of Langley Research Center); Lee R. Scherer, then Lunar Orbiter Program Director (later Director of Kennedy Space Center); and Leon J. Kosofsky, Lunar Orbiter program engineer. Complete chronological files of the Lunar Orbiter Program Office enabled me to outline the basic developments since the inception of Lunar Orbiter.

After studying files in Washington and at Langley Research Center and interviewing project officials, I went to Kennedy Space Center to witness the launch of Lunar Orbiter 5, the last mission of the program. There I interviewed program officials and Boeing and Eastman Kodak contractor representatives. Back in Washington, I wrote a preliminary manuscript about the program, for limited circulation among NASA offices as a Historical Note.

I returned to NASA Headquarters in the summers of 1968, 1969, and 1970 to expand my study of the program--one of NASA's major successes before the Apollo landings. In early June 1969, I was assigned to the Apollo Lunar Planning Office, whose director, Scherer, had encouraged me throughout the first two summers of research. In his office, I could see how Lunar Orbiter photographic data were being used in planning the Apollo 11 landing and subsequent missions. I conducted additional interviews and discussed results of Orbiter missions with Dr. Farouk El-Baz and Dennis James of Bellcomm, a consulting firm supporting NASA on Apollo. Through these talks I learned the technical and scientific significance of much of the Orbiter photography and how it was being applied. I went again to Langley, with new questions. Many of the former Lunar Orbiter project officials were occupied with a new planetary program: the Viking Program to explore Mars. Lunar Orbiter was history for them, but the experience from that program was already helping them in their newest endeavor. As this manuscript goes to press the two dual-role Viking spacecraft have successfully orbited Mars and sent two landers to the Martian surface. These craft have conducted numerous experiments to search for signs of life and to give us our first detailed views of the Martian landscape.

During the remainder of 1969 and in the summer of 1970 I worked to complete the draft of the history contained in the following pages. I submitted the manuscript in June 1971, shortly before beginning my present career as a Foreign Service officer.

The decade of the sixties was filled with turbulence, discontent, and upheaval. It also was a time of outstanding achievements in advancing our knowledge of the world in which we live. We accelerated the exploration of our planet from space. We landed men on the Moon, brought them safely home again, and learned how they could survive in space. And we began sending unmanned planetary explorers to chart the solar system and to search for signs of life on Mars. It is the purpose of this history to recount one chapter in this exploration, as a small contribution to the store of knowledge about America's first voyages on the new ocean of space.

I am grateful to the NASA History Office, whose staff have enabled me to write this history. I dedicate it to all the people who worked to make Lunar Orbiter the success it was--that they might have a record of their accomplishments to share with future generations.

Bruce K. Byers
Bombay, December 14, 1976

CHAPTER I

UNMANNED LUNAR EXPLORATION AND THE NEED FOR A LUNAR ORBITER

The Call for a Program of Exploration

During the decade of the sixties, three major ventures of the National Aeronautics and Space Administration thrust America's unmanned exploration of the Moon outside the Earth's atmosphere: the Ranger Program, the Surveyor Program, and the Lunar Orbiter Program. Initiated before President John F. Kennedy's May 25, 1961, request for a national decision to make a manned lunar landing in the sixties, Ranger and Surveyor gave the United States its first close look at the Moon. The original objectives of the programs had not envisioned imminent exploration of the Moon by men. Instead, NASA had developed highly proficient instrumented means for preliminary exploration without direct applications in an undertaking such as the Apollo manned lunar landing program.

One of the chief spokesmen for lunar exploration in the early days of America's space program was Nobel Laureate Harold C. Urey. In his address to the Lunar and Planetary Colloquium meeting on October 29, 1958, at the Jet Propulsion Laboratory, Urey called for a stepped-up United

States effort to explore Earth's natural satellite.[1] He summarized what scientists then knew about the origin and composition of the Moon: that much speculation but little conclusive knowledge existed concerning the Moon's environment.

Man had noticed many unique and unusual phenomena on the lunar surface through optical telescopes since Galileo's first observations in 1609, but Earth's atmosphere limited the explorative abilities of scientists. Urey concluded that automated probes would enable human observation to pierce the atmosphere for more detailed, precise looks at the Moon. Such probes would allow man to take the next logical step before actual manned lunar missions brought him to the Moon's environment and a landing on its alien surface. That surface, unlike Earth's, had not experienced millions of years of atmospheric erosion and weathering processes, as far as observations up to that time had revealed. What had it experienced? The answer to this question could possibly explain the birth and development of Earth and, indeed, of the solar system.[2]

---

[1] Harold C. Urey, "The Chemistry of the Moon," *Proceedings of the Lunar and Planetary Exploration Colloquium*, Jet Propulsion Laboratory, Pasadena, Calif., October 29, 1958, Vol. I, No. 3, pp. 1-9.

[2] *Ibid.*

Following Urey's call for intensified efforts to extend America's lunar exploration capabilities, but not necessarily in response to it, the newly created National Aeronautics and Space Administration requested the Jet Propulsion Laboratory to develop a study of the requirements for a multi-phase program to explore the Moon. Albert R. Hibbs, Chief of the Research Analysis Section at JPL, organized a study group to analyze the problem. On April 30, 1959, he submitted the group's findings to NASA Headquarters. Among other steps the Hibbs Report proposed placing a satellite

> in a well-controlled orbit around the moon using terminal guidance.... High resolution photographs of the surface of the moon will be taken at various wave lengths and polarizations. These photographs should provide information on the surface characteristics of the moon that will be valuable for choosing a site for a lunar soft landing.[3]

The Hibbs Report suggested a more sophisticated approach toward lunar exploration than that which NASA actually undertook, and it did not become the basis for the Lunar Orbiter Program. Nevertheless, it indicated the kind of probe which would perform necessary, extensive photography of the Moon's surface. The lunar orbiter con-

---

[3] Albert R. Hibbs (ed.), *Exploration of the Moon, the Planets, and Interplanetary Space*, JPL Report No. 30-1 (Pasadena, Calif.: Jet Propulsion Laboratory, California Institute of Technology, April 30, 1959), pp. 93-95.

cept later was adapted from the Surveyor Program which NASA Headquarters initiated with JPL in May 1960.

In December 1959 NASA and JPL had started the Ranger Program, the first step in NASA's unmanned lunar exploration venture. Surveyor, the second major program in this venture, originally envisioned two kinds of probes: a softlanding spacecraft for on-site investigation of the Moon's surface and an orbiter for investigation of the near-lunar environment. They would share common hardware, thereby probably reducing costs.

Both Surveyor Lander and Surveyor Orbiter, as Congressionally authorized programs, called for very sophisticated spacecraft whose hardware would require major development. The burden of this development fell upon JPL and together with the Ranger and Mariner programs made it the pioneering agency in the difficult process of designing and building automated, long-life spacecraft for deep space exploration.

The Surveyor Orbiter did not materialize. The Ranger and the Surveyor Lander programs, as first-generation spacecraft programs, came to overtax the manpower and facilities at the Jet Propulsion Laboratory, and the Centaur Rocket Program at the Marshall Space Flight Center experienced development problems and was eventually transferred to the Lewis Research Center. Centaur was to be the launch vehicle for Surveyor, and, as originally envisioned, it was to have a capability to put an 1,100-kilogram spacecraft into a translunar trajectory. At Lewis this capability was reduced to 950 kilograms, causing redesign of the Surveyor Lander.

In the wake of early Soviet space achievements the American space program became enveloped in far-reaching political competition with the Soviet Union. In this atmosphere the United States counted heavily on the Ranger and Surveyor programs, pioneering endeavors in the application of new technology, to achieve an urgently needed "first" in space.

The first six Ranger missions, between August 1961 and February 1964, experienced no complete mission success, but they acquired valuable data on the performance of systems. The publicity of their shortcomings heightened the tension, frustration, and anxiety among Americans about the state of U.S. technological prowess, while it drowned out the significance of the lessons learned by NASA and JPL. By June of 1964 the congressional Subcommittee on NASA Oversight had reviewed the Ranger Program and had concluded that

> ...progress in improving testing and fabrication techniques at JPL is a step-by-step process with little direction from NASA Headquarters and that major improvement actions take place primarily as a result of failures. The subcommittee recognizes that the Ranger Program is both unique and complex in the strictest sense of a scientific accomplishment and supervisory practices as currently

in use throughout the missile-space industry would go far to develop improved testing and fabrication procedures needed for a sophisticated spacecraft such as Ranger.[4]

Mustering for the Challenge of Space

Since its inception in 1958 the National Aeronautics and Space Administration had undertaken the development of new procedures in planning, organization and management, as well as in hardware fabrication and in training for mission operations. In 1964 Congress had found weaknesses in one of NASA's lunar programs that demonstrated clearly some of the difficulties which NASA had to overcome in the development of its program to explore the Moon. This long-range task greatly challenged the knowledge and the talent which America mustered, and the muster took place in a politically charged atmosphere in which the United States had decided to pit its scientific and technological resources and prestige against those of the Soviet Union.

The history of the Lunar Orbiter Program constitutes a significant chapter in the initial exploration of the Moon and in America's first decade in space. It is part

---

[4] Project Ranger, Report of the Subcommittee on NASA Oversight of the Committee on Science and Astronautics, U.S. House of Representatives, June 16, 1964, p. 23. Three of the first six Ranger missions were not completed because of malfunctions in the launch vehicles, not the spacecraft. Moreover, Ranger flew on NASA's first Atlas-Agena launch vehicle with all of the problems entailed in proving a new system. Finally, it is fair to state that the Mercury Program took priority over Ranger in the selection of Atlas rockets as launch vehicles.

of the record of the preliminary phase in the Apollo Manned Lunar Landing Program, and we must now turn to its origins for a closer study of its role in putting the first men on the Moon on July 20, 1969.

CHAPTER II

TOWARD A LIGHTWEIGHT LUNAR ORBITER

The Surveyor Program

As a major part of America's first lunar exploration effort NASA initiated the Surveyor Program in May 1960 with a dual objective: to build an unmanned lunar lander for surface investigations and to build a lunar orbiter for photographic coverage of the Moon, with instrumentation to explore and measure some of its environmental characteristics. Both would use the Atlas-Centaur launch vehicle. NASA charged JPL with the responsibility for carrying out the objectives of the Surveyor Program. JPL employed a conceptual philosophy for Surveyor which reflected the thinking of the Office of Space Sciences and which was similar to that of Ranger: design and build a common spacecraft bus to carry out different missions.[1]

On March 23, 1961, the Lunar Sciences Subcommittee of OSS recommended that an orbiter have the capability to: 1) achieve high-resolution photography which could define objects smaller than 10 meters in size, 2) obtain total photographic coverage of the limb area and of the far side of the Moon at a resolution of 1 kilometer,

---

[1] Transcript of Proceedings -- Discussion between Nicks, Milwitzky, Scherer, Rowsome, and members of the National Academy of Public Administration, NASA Headquarters, September 12, 1968.

3) take reconnaissance photographs of the lunar surface at 100 meters resolution, and, finally, 4) make stereo pairs of areas where high-resolution photography was planned.[2]

The idea of modifying the Surveyor Lander system to serve as an orbiter was very attractive to NASA Headquarters planners, but during the last quarter of 1961 the Office of Space Sciences began to review the feasibility of a Centaur-class orbiter in the weight range of 950 to 1,100 kilograms. On December 5 Charles P. Sonett, Chief of Lunar and Planetary Sciences at NASA Headquarters, requested his staff scientist Newton W. Cunningham to compile an inventory of JPL's programs and a description of their status.[3] Specifically he wanted to know the stage of development of the authorized Surveyor Orbiter.

Early in January 1962 Cunningham sent a report to Sonett detailing the activities which JPL had been conducting since 1958 pertaining to a lunar orbital mission. These amounted to the following: 1) a 1958 study on close photography of the Moon with a spacecraft launched by the Jupiter rocket, 2) the development of a unique camera system for _Pioneer IV_, 3) a study in 1959 for the Vega Program concerning instrumentation for a lunar probe in

---

[2] Memorandum from Newton W. Cunningham to Charles Sonett, NASA Headquarters, Washington, D.C., January 12, 1962, p. 6.
[3] _Ibid._

which a dual vidicon camera was to be used for obtaining low- and high-resolution photographs of the Moon,[4] and, finally, 4) a study made in 1960 of a lunar orbiter experiment.[5]

Cunningham also pointed out in his report that JPL scientists could not successfully adapt the Ranger photographic system for use in the Surveyor spacecraft and that no photographic system had been developed specifically for the long-life requirements of an orbiter mission. This was the general status of the Surveyor Orbiter at the beginning of 1962.

The advent of the Apollo Program soon changed the requirements for a lunar orbiter and placed urgent demands on the Office of Space Sciences for information on lunar surface conditions. Apollo needed these data in order to design hardware and missions, and in turning to the Office of Space Sciences the Office of Manned Space Flight helped to reshape the philosophy supporting the need for a lunar orbiter spacecraft.

Early Apollo Impact on Lunar Orbiter Planning

On June 15, 1962, the Office of Manned Space Flight submitted for the first time since the U.S. manned lunar

---

[4] Ibid., p. 2.

[5] Edwin F. Dobies, The Lunar Orbiter Photographic Experiment, Jet Propulsion Laboratory Section Report No. 1-48, June 1, 1960.

landing commitment a formal list of requirements to OSS for data on the Moon's surface. The list gave the Office of Lunar and Planetary Programs within OSS its first opportunity to compare the objectives of its lunar programs with preliminary Apollo needs. It re-examined the mission objectives of the Surveyor Lander and acknowledged that Ranger data would not meet the Apollo requirements.

It directed JPL to review all possible ways of converting the Ranger into an orbiter. JPL scientists and engineers soon responded that a conversion was not possible. JPL, in turn, requested the Hughes Aircraft Company, prime contractor for Surveyor, to examine the possibility of designing a 360-kilogram orbiter that the Atlas-Agena rocket could carry on a translunar trajectory. Hughes's report showed that such a lightweight spacecraft would have only a 27-kilogram payload, placing extreme constraints on the visual instrumentation system.[6] Following this up, JPL examined the feasibility of using the Agena with a Surveyor Kick Stage which would allow for a spacecraft weight of about 540 kilograms and a payload of 57 kilograms.[7]

---

[6] Support of Project Apollo by Programs in the Office of Space Sciences, Issue No. 1, July 30, 1962; Hughes Aircraft Company Document No. 262001, June 18, 1962.

[7] Ibid., p. 3.

However, this approach would require more research and development before NASA could pass judgment on its feasibility. Deciding that it did not have time to investigate this approach, the Office of Space Sciences proceeded with the Centaur-class Surveyor Orbiter.

By the end of July 1962 OSS had formulated the basic photographic requirements for the Surveyor Orbiter, but unfortunately these fell below the very demanding needs of Apollo. The Apollo Program required photographic data of the lunar surface that could show slopes of less than $7°$ with less than 1-meter protuberances and depressions on the surface of the Moon's front side. The first version of the Surveyor Orbiter would be able to shoot stereoscopic photographs of the lunar surface with a resolution only as small as 9 meters and monoscopic photographs which would resolve details as small as 1 meter. It would cover a minimum area of $100°$ longitude by $40°$ latitude from the equator on the visible side of the Moon.[8]

The spacecraft would most likely employ a television camera system. The Surveyor Orbiter photo system had one great drawback which the Support of Project Apollo document cited: "Landing area coverage of the size required [by Apollo] is not now possible except through repeated Ranger or Surveyor flights into the same area or by means

---

[8] Ibid., p. 7.

of a photographic roving vehicle or a hovering spacecraft."[9]

The level of technology in photographic systems for long-life lunar missions had not progressed much beyond the Ranger system, and NASA Headquarters recognition of this fact contrasted markedly with the status of the Surveyor Orbiter, on paper, as of July 20, 1962. Briefly summed up it was:

1. Five flights were planned.

2. Centaur rocket was to be launch vehicle; spacecraft weight was to be about 800 kilograms.

3. Jet Propulsion Laboratory was to establish design requirements and present them by September 1, 1962.

4. Surveyor Orbiter was to incorporate maximum amount of Surveyor Lander hardware and technology.

5. JPL was to develop a plan for the evaluation of experiments other than the Visual Instrumentation System by August 17, 1962. NASA Headquarters was to review this.

6. No Surveyor Orbiter Project Plan existed. JPL was to develop one and submit it to NASA for review by November 30, 1962.

7. A total of $29.5 million in funds existed for the Surveyor Orbiter in FY 1963 and $29.0 million in FY 1964. These funds would be redistributed between Surveyor Orbiter, Surveyor Lander, and the Ranger Improvement Plan only on the basis of defined relative values.[10]

The Jet Propulsion Laboratory had no operational Surveyor Orbiter program at this time. Indeed the troubles which

---

[9] Ibid., p. 8.

[10] NASA, Office of Space Sciences, Surveyor Orbiter Guidelines, July 20, 1962.

JPL was experiencing with the Ranger Program acted as a brake on the development of the orbiter.[11]

The Centaur Rocket Program

The Centaur Rocket Program did not facilitate JPL's work on Surveyor. The Marshall Space Flight Center, in charge of Centaur but with the Saturn Rocket Program as its prime responsibility, was experiencing development problems which caused the rocket's delivery schedule to slip, moving the earliest date for the first launch of a Surveyor Lander to late 1964. Moreover, the Centaur difficulties motivated officials in the Office of Space Sciences to review Surveyor Orbiter plans with the objective of obtaining an orbiter independent of Centaur. The Office of Space Sciences began to examine the idea of a spacecraft which might use existing hardware and the Agena rocket, already successfully tested in space. By September 1962 OSS had the requirements for, and the feasibility of, a lightweight lunar orbiter under serious study. Nevertheless, it had one major technological obstacle to surmount: developing a flexible, long-life photographic system capable of obtaining data to meet the requirements established by the Office of Manned Space Flight.

---

[11] Interview with Oran W. Nicks, Director of Lunar and Planetary Programs, Office of Space Science and Applications, NASA Headquarters, August 14, 1967.

The Search for a Lightweight Orbiter

On September 21 Oran W. Nicks, Director of Lunar and Planetary Programs in OSS, requested Lee R. Scherer a naval Captain on assignment to NASA, to form "a working group with appropriate representation from the Directorate of Lunar and Planetary Programs and consultants from other Headquarters offices, the scientific community and Field Centers...to study adaptations of the Ranger and Able 5 spacecraft to conduct lunar reconnaissance missions beginning in 1964...."[12] Nicks asked Scherer to confine his activity to the known spacecraft systems: the Ranger, the Able 5 built by Space Technology Laboratories (STL), and a system proposed by the Radio Corporation of America (RCA).

At the same time A.K. Thiel, Vice President in charge of Spacecraft Systems Program Management at STL, sent a detailed summary of a proposed lunar photographic satellite to Nicks at NASA Headquarters on September 20. The STL proposal offered for the first time a conceptual basis for a lightweight orbiter. It presented a plan for launching a spin-stabilized spacecraft into lunar orbit with the Atlas-Agena D. Once there the spacecraft's photographic system would take pictures of the Moon with a 254-centimeter

---

[12] Memorandum from Oran W. Nicks to Capt. Lee R. Scherer, OSS, September 21, 1962.

focal-length spin-scan camera very similar to one which Merton E. Davies of RAND Corporation developed in 1958.

The STL system did away with a cumbersome television payload and used a film system instead. Film had the definite advantage over television as far as its ability to obtain higher resolution photographs. Thiel stressed the reliability of the STL proposal and stated that his firm would be prepared to build and launch three spacecraft within 22 months from the go-ahead date.[13]

On October 15 Nicks informed Thiel that his office had the STL proposal under consideration. Meanwhile, within NASA discussion continued concerning the priorities in the American lunar exploration program.

OSS-OMSF Cooperative Planning

The Office of Space Sciences and the Office of Manned Space Flight soon discovered that in order to expedite a manned lunar landing before 1970 they had to define more precisely their working relationship and the Apollo requirements which unmanned lunar probes could fulfill. On October 23, 1962, Joseph F. Shea, Deputy Director of the Office of Manned Space Flight, informed Nicks that OMSF had confirmed "the relative priorities which should

---

[13] Letter from Dr. A.K. Thiel, Space Technology Laboratories, Inc., to Oran W. Nicks, Director, Lunar and Planetary Programs, OSS/NASA, Washington, D.C., September 20, 1962.

be attached to the development of unmanned lunar systems for acquisition of data on the lunar environment in support of the manned lunar program."[14]

Shea also informed Nicks that the Apollo Program had a more urgent need for the kind of data which a softlanding Surveyor could provide than for that which an orbiter could obtain in the near-lunar environment. The data which an orbiter could supply OMSF could directly apply to Apollo mission planning, but Surveyor data on the load-bearing conditions of the lunar surface had a more direct, immediate application in the design of the Lunar Excursion Module (LEM). Shea stressed that NASA should not commit itself to an orbiter in FY 1963 if this would jeopardize the present Ranger and Surveyor programs. This priority ordering from OMSF directly affected JPL's priorities with Surveyor.

In any case, Shea concluded, for an orbiter to provide the manned lunar landing program with useful data, it should concentrate on selenodetic and topographical conditions. This kind of data would permit the verification and selection of the initial sites for a manned lunar landing.[15]

---

[14] Memorandum from Joseph F. Shea, Office of Manned Space Flight, to Oran W. Nicks, Office of Space Sciences, October 23, 1962.

[15] Ibid.

Shea recommended to Nicks the establishment of a formal OSS-OMSF working relationship, and subsequently Homer E. Newell (Director, OSS) and D. Brainerd Holmes (Director, OMSF) announced the organization of the Joint OSS/OMSF Working Group with full-time representation from both offices. The group would be responsible for "recommending to OSS a program of data acquisition so as to assure the timely flow of environmental information into the planning for manned projects."[16]

While the Joint Working Group initiated greater cooperative efforts between the two NASA Headquarters offices, the work group which Nicks had requested Scherer to set up arrived at a decision on October 25 concerning its review of the studies for a lightweight orbiter. It recommended that the STL proposal be given more intensive consideration and that NASA drop RCA's proposal.[17] Several reasons supported the group's decision, and among them the Apollo requirements were the most important. As of November 16 these requirements stood as follows: An orbiter should be able to identify 1) 45-meter size objects over the entire surface of the Moon, 2) 4.5-meter objects in

---

[16] Memorandum for the Associate Administrator, NASA (Robert C. Seamans, Jr.), from Dr. Homer E. Newell, OSS, and D. Brainerd Holmes, OMSF, October 22, 1962, p. 1.

[17] Lee R. Scherer, Surveyor Program Engineer, Study of Agena-based Lunar Orbiters, NASA Headquarters, Office of Space Sciences, October 25, 1962, p. 1. See also Memorandum from Captain Lee R. Scherer to Oran W. Nicks, OSS, November 16, 1962, concerning STL Proposal No. SC5100 and Proposal No. SC5101.

the areas of prime interest, and 3) 1.2-meter objects in the landing areas.[18]

The Scherer Group's Report

According to the Scherer group, STL's orbiter seemed to have the greatest potential for fulfilling the requirements set by OMSF and OSS. The spacecraft would weigh about 320 kilograms, which placed it well within the Atlas-Agena launch vehicle capabilities. It would be spin-stabilized and its monopropellent propulsion system, capable of multi-starts, would give it the added flexibility of being able to change its orbital parameters around the Moon. This spacecraft could photograph the entire Moon from a polar orbit of 1,600 kilometers above the lunar surface and obtain pictures resolving objects as small as 18 meters across. If ground control placed the spacecraft in an equatorial orbit of 40-kilometer altitude, it could photograph the area along the lunar equator at the amazing resolution of 0.5 meter.[19] The Scherer group believed that these positive features of the STL system far outweighed the drawbacks involved in image motion compensation, the need for high-speed film, and for high shutter speeds in the camera.

---

[18] Scherer, *Study of Agena-based Lunar Orbiters*, p. 1.
[19] *Ibid.*, p. 2.

On the other hand the RCA approach, which the group rejected, consisted of injecting a 3-axis attitude-stabilized payload into lunar orbit from a Ranger-type bus. The photographic system onboard would employ a vidicon television which had two major weaknesses: low sensitivity in the vidicon unit and inadequate horizon scanners. In addition, the capsule that the Ranger bus would inject into orbit would weigh a mere 200 kilograms and this left little allowance for the actual payload hardware. The integration of the capsule and the Ranger bus and their separation before lunar orbit insertion further compounded the problem of weight limits on the payload. Even if this could be resolved with a high degree of reliability, the TV system could not detect objects smaller than 130 meters in wide-area coverage and 30 meters in limited-area coverage, at best.[20]

Scherer's group considered these negative aspects of RCA's proposal, together with the estimated cost of $20.4 million for building and flying only three spacecraft, too expensive and inadequate for the needs of Apollo. The group believed that pictures of the lunar surface of equal resolution could be obtained by far less expensive means,

---

[20] Ibid.

such as balloon-borne telescopes. The RCA proposal would require major research and development of a better visual instrumentation system in order to be capable of satisfying Apollo requirements, and this would be too costly in time and money.

There is irony in the Scherer group's final evaluation. The STL system won recommendation while the RCA system did not, and yet the final Lunar Orbiter spacecraft which NASA flew incorporated more of the concepts supporting the RCA system and less of those of the STL system. This was especially true of the attitude control system, although it did not apply for either of the camera systems.

Scherer's report to Nicks recommended that NASA fund two STL studies in 1963 in order "to better establish the feasibility of the proposed Able 5 lunar photographic spacecraft..." and "to provide more detailed information about the Able 5 spacecraft system and its photographic payload." The rationale for this decision was that it was "necessary to establish the confidence needed for duly considering a flight program of this type, should it be deemed preferable to a Centaur-based orbiter for any reason."[21]

Plans for the Centaur-based lunar orbiter began to

---

[21] Ibid., p. 1.

lose their attractiveness once Scherer's group had shown that an Agena-class orbiter, based upon STL research, would give NASA a more expedient means of data acquisition for Apollo requirements. Moreover, the status of the Centaur Rocket Program, originally managed by the Marshall Space Flight Center and then transferred to the Lewis Research Center, did not make the concept of a Surveyor Orbiter more acceptable. Flaws in the rocket's basic fuel tank configuration and delays in the development tests eventually influenced the schedules of the Surveyor Lander at JPL because the overall capability of the Centaur was reduced from 1,100 to 950 kilograms.[22]

Problems at JPL

The Jet Propulsion Laboratory was encountering increasing problems with the Ranger Program which further influenced the progress of the Surveyor Program. The problems and the added pressure of the Apollo Program's newly introduced priorities gave increased support to the move to define and establish criteria for an Agena-class lunar orbiter program within the Office of Lunar and Planetary Programs.

---

[22] Memorandum, Dr. Homer E. Newell, Office of Space Sciences, NASA Headquarters, November 1, 1962. (Joseph Ziemanski, former Agena Project Engineer, Lewis Research Center comments that the Lewis Research Center met its scheduled delivery date with the first Centaur in the Surveyor Program, but no Surveyor was ready to be launched on the original launch date.)

In pursuit of his responsibilities with the authorized Surveyor Orbiter and without the knowledge of the Scherer group's findings, Clifford I. Cummings, JPL Lunar Program Director, informed Oran W. Nicks on October 26 that JPL was planning to undertake another study of the Surveyor Orbiter and its mission. He stated that JPL desired to spend $1.5 million of its FY 1963 budget to do this work, and he included in his memorandum to Nicks a proposed plan of study for a lunar orbiter spacecraft.[23]

Nicks immediately answered the JPL request with a letter to Cummings in which he outlined the numerous study efforts already performed or in the process of completion. He pointed out the concern of NASA Headquarters about the growing disparity between the status of the Surveyor Program at JPL and that of the Centaur Program. He informed him that Headquarters had already proceeded to examine the feasibility of an Agena-class orbiter. Thus an additional study would not serve.

The difficulties encountered in the first four Ranger missions in 1961 and 1962 and the great effort made to

---

[23] Memorandum from Clifford I. Cummings, Director of Lunar Programs, JPL, to Oran W. Nicks, Director, Office of Lunar and Planetary Programs, NASA Headquarters, October 26, 1962, and memorandum in reply from Oran W. Nicks to Clifford I. Cummings, November 8, 1962, p. 2. See also Brief History of Lunar Orbiter Work, prepared for Edgar M. Cortright, NASA Headquarters, May 2, 1963.

obtain a launch vehicle which Lunar Orbiter would later use kept the Jet Propulsion Laboratory totally committed to the Ranger and Surveyor Programs. NASA Headquarters, meanwhile, approached Floyd L. Thompson, Director of the Langley Research Center, early in 1963 about the possibility of taking on a lunar orbiter project.

Langley Enters the Picture

On January 2, 1963, while attending a Senior Council Meeting of the Office of Space Sciences at Cape Canaveral, Floyd L. Thompson met with Oran W. Nicks, who asked him if the Langley Research Center would be willing to study the feasibility of undertaking a lunar photography project. The Langley Director agreed to have his staff study the project.[24]

Nicks had suggested to senior staff members within OSS the idea of approaching the Langley Research Center about a possible lunar orbiter project for several reasons. First, JPL had more than enough to accomplish with Ranger and Surveyor. Its manpower and management capabilities could be stretched only so far. Secondly, the Langley Research Center, founded in 1917 to develop an aeronautical

---

[24] Memorandum from Floyd L. Thompson, Langley Research Center, to Dr. Eugene M. Emme, NASA Historian, NASA Headquarters, Subject: Comments on draft of Lunar Orbiter History dated November 4, 1969, December 22, 1969.

research capability for the United States, had proved itself to be very successful in project management. Finally, a wider distribution of operational programs among NASA field centers appeared to Nicks to be a prudent management decision, allowing the centers to develop new and varied capabilities for future NASA ventures.[25]

Langley put forth an intensive effort and by March 1963 completed its assessment of the task of obtaining the required lunar photography and of its capability to manage a lunar orbiter project.

In the fall of 1962 Nicks had requested Lee Scherer and Eugene Shoemaker, a geologist on loan to NASA from the United States Geological Survey, to define more exactly the Apollo requirements for photographic data which an orbiter could best satisfy. The two men spent the remainder of the year and early 1963 examining Ranger and Surveyor spacecraft components which might be best used in a lightweight orbiter. Concurrently Dennis James of Bellcomm, a private research and advisory organization working with the Office of Manned Space Flight, conducted another review of existing technology and hardware which might be usable in a lunar orbiter.

---

[25] Interview with Oran W. Nicks, NASA Headquarters, August 14, 1967.

In October 1962 the Office of Space Sciences had followed up the recommendation of the first Scherer group in a further move to define the requirements for an Agena-class orbiter and had let a contract to the Space Technology Laboratories to "make a detailed preliminary study of a spin-stabilized lunar photographic spacecraft based upon the Able 5 development to be launched by the Atlas-Agena vehicle."[26]

STL conducted the study, and during a major planning and review meeting at the Langley Research Center on February 25, 1963, representatives from OSS, OMSF, Bellcomm, STL, and Langley reviewed the preliminary conclusions of the STL research. Following this meeting both Langley and NASA Headquarters stepped up their activities to formulate a viable basis for an Agena-class orbiter.

Space Technology Laboratories continued to work on a reliability assessment of a lunar orbiter photographic mission and analyzed the problem of having a lunar orbiter locate and photograph a landed Surveyor. Dennis James of Bellcomm developed a study for Joseph F. Shea of OMSF and Lee R. Scherer of OSS concerning the role a lunar orbiter could play in the manned and unmanned exploration

---

[26] Project Approval Document dated October 16, 1962, drawn up by Captain Lee R. Scherer, Office of Space Sciences.

of the Moon.[27]

Langley personnel continued to study the feasibility of a lightweight orbiter during the remainder of February. Their activity was independent of the STL study and, on March 5 at a second plenary meeting at Langley, representatives from STL and Langley presented the findings of their two studies to officials from OMSF, OSS, Langley, and Bellcomm.[28]

Amazingly the two independent analyses came to very similar conclusions. First, the probability factor of one mission success out of five attempts was approximately 93/100, based upon known systems. The probability of two successes in five was about 81/100. In addition the studies confirmed that an orbiter using existing hardware could photograph a landed Surveyor and thus definitely assist in Apollo site verification. On the basis of these data the members of the meeting concurred that an unmanned lunar orbiter had an extremely important role to play in the pre-Apollo phase of the Moon's exploration.[29] The next major step was to convince top Headquarters management

---

[27] Status Report on Orbiter -- Thursday, February 28, 1963, from the Director, Lunar and Planetary Programs, to the Assistant to the Director for Manned Space Flight Support.

[28] Memorandum from Homer E. Newell, Director, Office of Space Sciences, to the Director, Office of Space Flight, concerning questions on unmanned lunar orbiter, March 14, 1963.

[29] Letter from Dr. Floyd L. Thompson, Director, Langley Research Center, to NASA Headquarters -- Code SL, attn. Scherer, March 6, 1963.

that an Agena-class orbiter could best accomplish exploration for both the Office of Space Sciences and the Office of Manned Space Flight. To this task OSS and Langley now turned.

Following the March 5 meeting at Langley, Floyd Thompson's staff made a presentation of Langley's assessment at NASA Headquarters to Associate Administrator Robert Seamans, Jr. Clinton E. Brown acted as spokesman for the center and presented the following basic points to Dr. Seamans and members of the Office of Space Sciences:

1. Langley had the capability to handle a lunar orbiter project, but it would require an additional 100 persons if it was to avoid serious interference with its commitments to the Office of Advanced Research and Technology.

2. Analyses showed that it was feasible to obtain the desired lunar photography.

3. The contract for the project should be made on a competitive basis despite the work which STL had conducted on a preliminary Agena-class lunar orbiter system.[30]

Establishing Management Arrangements

The Office of Lunar and Planetary Programs within the Office of Space Sciences acted as coordinator of the various activities required by a new lunar orbiter program. Langley, once it had assessed its ability to undertake a

---

[30] Memorandum, Thompson to Emme, December 22, 1969, p. 2.

major unmanned deep space project to obtain lunar photography, began to develop formal plans for conducting such a project. It used the guidelines established in General Management Instruction 4-1-1, effective as of March 8, 1963.

General Management Instruction 4-1-1 covered planning and implementation of NASA projects and was part of an agency-wide management reform which NASA Administrator James E. Webb had initiated in October 1962. GMI 4-1-1 specifically "prescribes the policies and procedures for project management within NASA with respect to the manner in which projects are planned, approved and implemented."[31] These applied to NASA Headquarters, the field centers, and JPL.

Under GMI 4-1-1 a program was defined as "a related series of undertakings which continue over a period of time (normally years), and which are designed to accomplish a broad scientific or technical goal in the NASA Long-Range Plan; e.g., Lunar and Planetary Exploration...."[32] The appropriate Program Office (i.e., Office of Space Sciences) had the responsibility of carrying out the program. Supporting the program activity was the project, which, within a

---

[31] NASA Management Manual, Part I, General Management Instructions, Chapter 4, Number 4-1-1, March 8, 1963, p. 1 (hereinafter cited as GMI 4-1-1).

[32] Ibid.

program, was "an undertaking with a scheduled beginning and ending...."[33]

Within the project was the system -- "one of the principal functioning entities comprising the project hardware within a project or program." The system consisted of a number of subsystems, each a functional entity within it. Lunar Orbiter was such a system.[34]

The GMI 4-1-1 established four basic policies applicable to a program: 1) Project Initiation, 2) Project Approval, 3) Project Implementation, and 4) Organization for Project Management. Of these the second required that for any given project a Project Approval Document (PAD) be drawn up. This document would give a brief description of the proposed project's scope, of its assignment and its system management responsibility, and of the resource requirements by fiscal year. The Associate Administrator of NASA (in this case Seamans) had to approve the PAD before any steps to implement the project could be taken.[35]

Once the Associate Administrator had signed the PAD, the third policy came into effect. The first major step in implementing a new project was the drafting of the Project Development Plan (PDP), which the respective

---

[33] Ibid.
[34] Ibid.
[35] Ibid., p. 4.

Program Director (in this case Homer E. Newell, Director of the Office of Space Sciences) had to approve. The PDP had to describe in specific terms the technical, financial, procurement, and management arrangements for the project. It had to state clearly the assignment of managerial responsibilities and authority, manpower, and facilities and the procedure for funding.[36]

Finally the fourth policy stated that "the organizational pattern for a given project to system will be determined on a case-by-case basis. The centers or Headquarters Offices having project and system management responsibilities will be described in the Project Approval Document approved by the Associate Administrator. The detailed assignment of responsibility and authority will be described in the Project Development Plan."[37]

The policy of Organization for Project Management also established the roles which Headquarters and the field centers would play in a given project. Headquarters held the following specific responsibilities:

1. Establishment of objectives and policy guidelines.
2. Allocation of resources and provision for reprogramming.

---

[36] Ibid., pp. 4-5.
[37] Ibid., p. 5.

3. Provision of decisions and resources not within the scope of approved Project Development Plan or not otherwise within the field center authority.

4. Performance of inter-project coordination.

5. Evaluation of overall performance and accomplishment of project objectives.[38]

The brief, foregoing explanation of GMI 4-1-1 will enable the reader to assess how Langley went about preparing for the Lunar Orbiter Program during the course of 1963 up to August 30. During March the Langley Research Center formulated a Project Approval Document for a lightweight orbiter. It was assisted by Scherer and Shoemaker at NASA Headquarters and by the studies which STL and Bellcomm had conducted.

On March 25, 1963, the Project Approval Document was finished. Floyd L. Thompson and Sherwood L. Butler, the Langley Contracting Officer, submitted it to Associate Administrator Robert C. Seamans, Jr., together with a procurement document on this date. At the same time Langley also finished drafting a preliminary Project Development Plan, which it sent to Deputy Associate Administrator, Office of Space Sciences, Homer E. Newell at the end of March.[39]

---

[38] Ibid., p. 6.

[39] Project Development Plan for Lunar Orbiter Project (updated December 1964 and June 10, 1966), Langley Research Center, Project No. 814-00-00. p. II-2.

The Office of Space Sciences faced several major management decisions at this time which influenced the initiation of a new lunar orbiter program. Among these OSS had to decide what action to take on a lunar orbiter in the face of a projected shortage of funds in FY 1964. At the time that OSS submitted its FY 1965 budget estimates, it held that the initiation of a new orbiter project was not financially realistic.[40]

However, Langley's quick assessment of its ability to take on the orbiter project enabled the Deputy Director of OSS, Edgar M. Cortright, to recommend to OSS Director Homer E. Newell that it be initiated. Cortright's recommendation was not based only on Langley's assessment. Following the submission of the FY 1965 budget estimates his office received new information which made it more feasible to decide on a start for a new lunar orbiter project.

First, the Office of Manned Space Flight had endorsed the orbiter, and OSS had made a tentative analysis of its ability to meet the needs of the manned program. Secondly, Cortright had assessed through numerous meetings with people from OSS, OMSF, JPL, and the Goddard Space Flight

---

[40] Memorandum from SD/Deputy Director, OSS, to S/Director, OSS, concerning: Recommended reprogramming within the Office of Space Sciences, April 25, 1963.

Center (GSFC) that an orbiter project was definitely needed and feasible.[41]

He outlined to Newell the major factors to be considered in the lunar orbiter decision:

1. The STL-type lunar orbiter had been studied by OSS, OMSF, Bellcomm, and LRC and had been found to be feasible and desirable.

2. One successful orbiter would be worth dozens of successful Ranger TV impacters.

3. Langley could provide the management within its present ceiling, if necessary, and was highly motivated to do so.

4. The orbiter would be a new start and would probably have its share of unforeseen problems. The technology was not quite "off-the-shelf" and the schedule for a 1965 launch would be tight.

5. The Apollo Program might plan a photo-reconnaissance mission capability.[42]

In view of these and other decisions pending on the Ranger program extension and the Mariner B flight, Cortright concluded that the Office of Space Sciences should "initiate the lunar orbiter project at 1.7 million in FY 1963, and 27.9 million in FY 1964. Contract award would await Congressional action on FY 1964 funds. Retreat is therefore possible."[43] A new start could be absorbed if the Block V Ranger were dropped. (Cortright recommended

---

[41] Ibid.

[42] Ibid., p. 2.

[43] Ibid., p. 4.

that it and subsequent Ranger blocks be dropped.) The $99 million programmed for Ranger would more than cover orbiter needs in FY 1965 since they would be about $71 million.[44]

Langley Develops the Request for Proposal Document

The approval of the Project Development Plan set the stage for drafting the Request for Proposal document (RFP) with which NASA would go to the aerospace industry in search of a contractor for Lunar Orbiter.

Of the assignments made in the PDP, the Langley Research Center (LRC) was to handle the project management and spacecraft system management responsibilities for Lunar Orbiter. In addition it had charge of overall project-wide systems integration between the spacecraft and the launch vehicle and the spacecraft ground support facilities, including communications, tracking, and data-acquisition systems.[45]

The Project Development Plan assigned to the Director of LRC overall technical, operational, and financial management for the Lunar Orbiter Project. In turn the Director was to implement project management through the

---

[44] Ibid., p. 5.

[45] Project Development Plan, Appendix, Attachment 1, pp. XII-1, XII-2.

Project Manager (Clifford H. Nelson). The Project Manager, working with a team of men, each expert in a specific area of the project, exercised control over plans, schedules, costs, technical changes, and data in order to obtain the most advanced lunar photographic and selenodetic information as early as possible.[46]

During the spring of 1963 Bellcomm continued to define lunar orbiter objectives for the Office of Manned Space Flight. Early in May it informed Scherer in OSS that "there are at the moment no fully developed lunar orbiter systems."[47] Subsequently it submitted a document entitled "Orbiter Recommendations" to Scherer. He reviewed it and forwarded it to Clinton E. Brown at Langley with the statement that, "although specific recommendations are subject to change on review by the Office of Space Sciences, it is considered an excellent document for guidance of Langley Research Center in preparation of the Request for Proposal for the Lunar Orbiter."[48]

The Bellcomm and Scherer groups assisted Langley in the work on the RFP while, at the same time, Oran W. Nicks briefed Dr. Robert C. Seamans, Jr., on the initiation of

---

[46] Ibid., Appendix, Attachment 2, p. XII-3.

[47] Bellcomm Working Paper, submitted by W.S. Boyle to J.F. Shea, May 10, 1963, p. 3; Bellcomm study on lunar orbiter objectives, May 14, 1963.

[48] Letter from Capt. Lee R. Scherer, NASA Headquarters, to Clinton E. Brown, Langley Research Center, May 24, 1963.

the new lunar orbiter and its impact on the Block V Ranger series of spacecraft.[49]

In a further move to assist Langley in drafting the RFP, the Office of Manned Space Flight submitted a revised summary of the Apollo requirements to OSS. It stated these critical needs: 1) data on radiation flux over a typical two-week period, 2) a summary and analysis of all efforts for short-term prediction of severe solar proton events, 3) measurements of particles capable of penetrating 0.01-centimeter and 0.1-centimeter aluminum during an average and a peak two-week period of micrometeoroid activity, and 4) photographic data on lunar surface conditions capable of showing cones 3.5 meters high and slopes of 15° inclination in an area of 60-meter radius, before the fall of 1965, and thereafter equivalent data showing cones 50 centimeters in height and slopes inclined 8° in an area of 1,600-meter radius.[50]

Other major needs were: 1) the measurement of the distribution of slopes greater than 15° in areas 7 meters in diameter; 2) photographs of at least 25-meter resolution over the largest possible area within ± 10° latitude and

---

[49] Memorandum from Edgar M. Cortright for Messrs. Nicks, Cunningham, Kochendorfer, Mitchell, Subject: Briefing of Seamans on current program proposals, May 15, 1963.

[50] Summary of OMSF Data Requirements Document, no date. See also: Discussion of Lunar Surface Photographic Requirements, Appendix III, April 19, 1963.

0° to 60° west longitude on the Moon.[51]

While the Office of Manned Space Flight and the Office of Space Sciences coordinated their activities through the Joint Working Group, officials at the Langley Research Center prepared the Request for Proposal document and the requirements of a lunar orbiter contract. NASA Headquarters representatives met with Dr. Thompson and his staff at Langley on June 25 to reach an agreement on the type of contract to be utilized in the procurement of the Agena-class lunar orbiter spacecraft.

Headquarters took the position that the contract should employ a cost-plus-incentive-fee mechanism similar to that used in the Pioneer Program. Langley officials, on the other hand, desired the cost-plus-fixed-fee contract because they expected unknown development problems to arise. They felt that such a contract would be easier to administer in that case. Headquarters officials remained vague about the nature of incentives which should be incorporated into the contract.[52]

Langley officials concerned with the determination of the kind of contract to be used remained firm on the point

---

[51] Ibid.

[52] Office of Space Sciences, memorandum to SL Files from SL/Assistant to the Director for Manned Space Flight Support, Subject: Meeting on Incentive Contracting for Lunar Orbiter at Langley Research Center, June 25, 26, 1963.

of retaining sufficient flexibility in seeking a contractor and negotiating a contract that would best suit Langley's needs. Thompson insisted from the beginning that all bidding be competitive. He was not convinced that Space Technology Laboratories had a decided advantage over other firms in the field, despite STL's research on lunar orbiter systems. He also made it clear that Langley would not commence work with a contractor under a Letter of Intent. Instead the contract would have to be negotiated and signed, and it would have to reflect, as closely as possible, the actual work it entailed. This would eliminate any basis for defining the nature of assignments following the initiation of work.

NASA Headquarters officials favored a spin-stabilized spacecraft and desired that the RFP reflect a preference for this kind of system. However, Langley officials insisted that they not be frozen to one concept for a spacecraft system. They wanted to see what exactly the aerospace industry could produce before selecting the spin-stabilized system. Although NASA's research into a lightweight orbiter had shown that the spin-stabilized system was feasible, Langley wanted room left for an attitude-stabilized (three-axis-stabilized) spacecraft system.[53]

---

[53] Interview with Floyd L. Thompson, former Director of the Langley Research Center, NASA Headquarters, January 29, 1970.

The June 25 meeting at Langley resulted in a compromise solution which would use the cost-plus-incentive-fee contract for procurement. Preliminary incentives were also established, but room was left for further suggestions from potential bidders.

Following this Homer E. Newell, Director of the Office of Space Sciences, sent a statement to Floyd L. Thompson at Langley on July 1 in which he further clarified the Headquarters position on Lunar Orbiter and its objectives. Thompson had expressed concern that the proposed orbiter project might be greater and more sophisticated than Langley had first estimated. Newell explained that his office maintained a policy of giving the needs of the Office of Manned Space Flight maximum support as far as such support did not impinge on OSS goals. At that time, Newell explained, the OSS specifications for a lunar orbiter could be approached but not entirely reached by an Agena-class orbiter. The Bellcomm studies had developed objectives for a lunar orbiter which would not fully satisfy Apollo requirements. Bellcomm's review and the STL proposal showed that these objectives represented the limits of feasibility up to that time.[54]

---

[54] Memorandum from Dr. Homer E. Newell, Director of the Office of Space Sciences, to Dr. Floyd L. Thompson, Director of the Langley Research Center, July 1, 1963.

Newell assured Thompson that although the proposed high-resolution photography, capable of pinpointing a landed Surveyor, seemed to be beyond feasibility, Langley did not have to rely upon the Bellcomm work to reach a decision. It could use the Bellcomm studies merely as a reference for determining the kind of Agena-class orbiter which could best accomplish the objectives of providing OMSF-Apollo with the data it required. If this were too impractical for Thompson, then Newell was open for any alternative suggestions.[55]

During July Langley and NASA Headquarters worked closely on the Request for Proposals. Headquarters desired that the RFP indicate to bidders that NASA was going to insist upon a very close working relationship with the contractor in selecting and approving subcontractors for the photographic data-acquisition components. NASA would reserve the right to determine the selection of the manufacturer of the sensor in the spacecraft system in order to obtain the best sensor regardless of any relationship between the prime contractor and the subcontractors.[56]

OSS officials desired that the Statement of Work, accompanying the RFP, indicate that NASA favored a spin-

---

[55] Ibid.

[56] Headquarters Comments on Documents for the RFP of the Agena-class lunar orbiter, no date, p. 1

stabilized spacecraft. Despite the recognition that such a spacecraft was feasible, simpler and less expensive than an attitude-stabilized system, Langley argued that the Request for Proposals should also allow bidders to offer an attitude-stabilized spacecraft. It was a sound argument. Langley would have the responsibility for the spacecraft system, and it wanted to explore all possible concepts. A compromise agreement was reached, providing that if bidders could offer approaches which differed from the established specifications but which would result in substantial gains in the probability of mission success, reliability, schedule, and economy, then NASA certainly invited them to submit such alternatives.[57]

Stipulations of the Request for Proposal Document

NASA Headquarters and Langley agreed that the RFP should explicitly clarify that the main mission of the new lunar orbiter was the acquisition of photographic data of high and medium resolution for selection of suitable Apollo and Surveyor landing sites. The secondary objectives provided for the acquisition of information about the size and shape of the Moon and about the properties of its gravitational field. The orbiter would also measure certain other lunar environmental characteristics in the Moon's vicinity.

---

[57] Ibid., p. 2.

However, the RFP was to state clearly that under no circumstances would these secondary objectives be allowed to dilute the major photo-reconnaissance mission. For this reason the Statement of Work which was to accompany the RFP was not to give any detailed descriptions of the secondary objectives.

In outlining the photographic requirements which the RFP was to make explicit, NASA Headquarters counseled Langley to use the following guidelines for identifying cones and slopes on the lunar surface. Cones were assumed to be circular features at right angles to a flat surface. These could be considered as recognized if the standard deviation of the cone's estimated height caused by system noise in the spacecraft was less than 1/5 of the cone's height. Slopes were assumed to be circular areas inclined with respect to the plane perpendicular to local gravity. Again a slope would be considered as recognized if the standard deviation of estimated slope caused by system noise was less than 1/5 of the slope.[58] These criteria required at least two photographic modes in the orbiter to obtain the data: 1) high resolution of limited areas and 2) wide coverage at medium resolution. Any bidder's

---

[58] Ibid., pp. 7-8.

proposal had to meet this requirement. However, a proposal would not have to employ both modes of photography on any one mission.

The Request for Proposals had also to state clearly that a bidder would provide in his proposal for instrumentation and telemetry capable of measuring certain characteristics of the lunar environment. These components would have to function independently of the photographic subsystem in order to record data regardless of the success or failure in obtaining pictures. Among the various environmental conditions which might be measured, micrometeoroid flux and total exposure to energetic particles and gamma radiation were two whose measurement would be necessary for gauging the performance of the spacecraft while also providing vital data for the Apollo Program.

In addition to this instrumentation the bidder would have to be able to determine precisely the altitude of his spacecraft at the time of each photographic exposure, the orientation of the picture in relation to lunar north, and the relative angle of the Sun to the portion of the Moon's surface covered by any photograph. The bidder would have to demonstrate his capability for providing such data as would be necessary to position all points within an area of contiguous coverage while being able to pinpoint 90% of all well-defined points to within 100 meters of their true horizontal positions relative to each other in the high-resolution mode. Finally the RFP was to require each bidder to be able to give the locations of photographed areas within one kilometer of their correct positions in the lunar system.[59]

Headquarters defined what it desired that the RFP do on the basis of the STL and Bellcomm studies, with the results of the two Scherer groups' research. Thus the spin-stabilized spacecraft system was preferable to Headquarters, but the RFP, in final form, did not precisely state which kind of spacecraft system would best do the job.

By August 1 Langley was concluding its preparations on the RFP. It also had drawn up the Statement of Work (SOW) document to accompany the RFP when it was released. The SOW set forth explicit guidelines for each bidder to use in developing a proposal. In addition to a general description of the mission which Lunar Orbiter would perform, the document stated the requirements which the spacecraft system would have to fulfill, the testing procedures and the interfaces which the contractor would have to establish and carry out, and the division of tasks which the contractor would have to perform.[60]

---

[59] Ibid., pp. 11-12.

[60] Statement of Work, Lunar Orbiter Project, Langley Research Center, March 18, 1964, Exhibit A.

Langley reached an understanding with Headquarters on the contract, which was to have incentives based upon cost, delivery, and performance.[61] Late in August Scherer presented a summary of Langley's Request for Proposal document to Nicks and Cortright, and on August 30, 1963, after Dr. Robert C. Seamans, Jr., had reviewed the RFP, NASA released it to the potential bidders. This step officially initiated the Lunar Orbiter Program.[62]

---

[61] Status Report on Lunar Orbiter, Langley Research Center, August 1, 1963.

[62] Letter from Capt. Lee R. Scherer to Oran W. Nicks and Edgar M. Cortright, Office of Space Sciences, NASA Headquarters, Washington, D.C., August 23, 1963.

CHAPTER III

BEGINNING THE LUNAR ORBITER PROGRAM

Congress Questions NASA on Orbiter

NASA's new Lunar Orbiter Program began while Congress was conducting annual authorization hearings. During August 1963 top NASA officials waged an impressive fight for more funds for an orbiter. They had to answer queries from the House Committee on Appropriations concerning their move to initiate a new orbiter project when the Surveyor Orbiter Project already had authorization and funds. The Committee claimed that NASA had channeled much of the money into other projects and that this attested to their higher priorities. Almost nothing had been spent on the Surveyor Orbiter.[1] The Committee seemed to think that NASA's lack of progress on its original concept of the Surveyor Orbiter and its development of a new lunar orbiter concept for a different project at Langley meant that it did not consider the mission of an orbiter as important as it wished Congress to believe.

Seamans, Dryden, Newell, and Cortright from NASA

---

[1] *Independent Offices Appropriations for 1964*, Hearings before a Subcommittee of the Committee on Appropriations, House of Representatives, 88th Congress, first session, August 19-20, 1963, p. 412.

Headquarters, and Pickering from JPL all provided testimony to clarify NASA's position on the Surveyor Orbiter and the urgent need for a lightweight lunar orbiter which could obtain vital data for the Surveyor Lander and Apollo programs. After their testimony before the Senate Committee on Aeronautical and Space Sciences, the Senate restored the proposed $28.2 million in funds for FY 1964 for an orbiter which the House had deleted from its authorization bill. Both houses reached a compromise late in August and authorized a total of $20.0 million for an orbiter.[2]

Appropriation hearings pertaining to the lunar orbiter project were scheduled to begin on October 18, but the Office of Space Sciences relied upon the approved authorization as a reasonable assurance that funds would not evaporate after the Lunar Orbiter Program was under way.

The Lunar Orbiter Project Office Is Established

With the Request for Proposals already sent out, the fledgling Lunar Orbiter Project Office (LOPO), under the direction of Clifford H. Nelson, set up shop at the end of August in the Langley Research Center's sixteen-foot wind

---

[2] House of Representatives, NASA Authorization for Fiscal Year 1964, Conference Report (to accompany H.R. 7500), House Report No. 706, August 26, 1963, p. 1.

tunnel facility in the West Area. The members of the original LOPO nucleus included Israel Taback, Robert Girouard, William I. Watson, Gerald Brewer, John B. Graham, Eugene A. Brummer, Robert Fairbairn, and Anna Plott, the last conducting all secretarial tasks. William J. Boyer joined the group soon after its formation.

Langley Center Director Floyd L. Thompson was instrumental in selecting Nelson as Project Manager. Very experienced at Langley, Nelson had the technical skills and the ability to work closely with people which his assignment required. Ideally a project manager should be capable of serving all vital managerial functions in a project. These include business as well as technical responsibilities. Nelson met most of the requirements which these responsibilities entailed.

Dr. Thompson brought James S. Martin, Senior Engineer at Republic Aviation, into Langley in October 1964 to assist Nelson in the realm of business management for the project. Coming from the aerospace industry to NASA, Martin had extensive experience in handling the business problems of contractors, and he was very capable of getting a job done. He had great knowledge of the operations of industrial contractors, something which Nelson and his staff needed. Martin's area of competency complemented that of Nelson and the two men formed a good team.

Both successfully instilled in the other members of the Lunar Orbiter Project Office a sense that the whole venture depended upon their individual work. Each member of the team came to see how his job fitted into the overall objectives of the project. Dr. Thompson assisted Nelson and Martin in the task of establishing good working relationships among those divisions at LRC which would lend support to Lunar Orbiter and among the other NASA and contractor personnel who had a part in the program.

Preparing for Contract Bids

At NASA Headquarters Lee R. Scherer, the Lunar Orbiter Program Manager, issued a status report to Oran W. Nicks and Homer E. Newell on September 4, stating that Seamans had signed the Project Approval Document on August 30. It called for five flight spacecraft using the Atlas-Agena D launch vehicle. The program would rely on the tracking and data-acquisition facilities of the Jet Propulsion Laboratory and the Deep Space Network which JPL was under contract to NASA to operate. The Deep Space Network (DSN) consisted of the Deep Space Instrumentation Facility (DSIF) and the Space Flight Operations Facility (SFOF). Langley had the responsibility to establish interfaces between its Project Office and those offices at these facilities which would assist the Lunar Orbiter Program.[3]

---

[3] Lunar Orbiter Status Report, OSS Review, September 4, 1963.

NASA's decision to build a new lunar orbiter attracted several aerospace firms engaged in research and development for America's space exploration effort. While Congress questioned NASA and the Office of Space Sciences continued planning, five major aerospace companies began to develop proposals in the hope of submitting the winning bid for the new spacecraft.

In Aviation Week & Space Technology, a major aerospace periodical, Richard G. O'Lone briefly surveyed the nature of NASA's Lunar Orbiter contract. He stated that the Lunar Orbiter Program was to be "the first major National Aeronautics and Space Administration project that will include cost, delivery and technical performance incentives as part of its contract."[4] O'Lone stressed that "selection of the orbiter as its first major incentive venture illustrates the urgency NASA attaches to the program."[5] In addition NASA included substantial incentives based upon predetermined rates for all underruns and penalties for overruns on deadlines. These it had made explicit so that the contractor would know the limits within which he could work.

However, NASA officials were quick to state that the

---

[4] Richard G. O'Lone, "Orbiter Is First Big NASA Incentive Job," Aviation Week & Space Technology, Vol. 79, No. 15 (October 7, 1963), p. 32.

[5] Ibid.

Lunar Orbiter incentive contract did not "mean that NASA has shifted its emphasis from a firm's technical management ability to the price it quotes for a job."[6] More significantly for Lunar Orbiter, "incentive contracting compels both NASA and the contractor to define what they want at the earliest practical date."[7] This had been Langley's major intention with the Request for Proposal document, and the aerospace companies bidding for the contract had to reflect in their proposals a well-defined understanding of the RFP.

While the potential contractors developed proposals for a lunar orbiter spacecraft, NASA's Office of Lunar and Planetary Programs accelerated its planning for the new lunar exploration venture at Headquarters. The Langley LOPO did likewise. Oran W. Nicks met with Floyd L. Thompson, Clinton E. Brown, Clifford H. Nelson, Charles Donlan, Eugene Draley, and Harold Maxwell at the Langley Research Center for a management conference on Tuesday, September 11, to discuss at length the major management aspects of the program. Lee R. Scherer and Leon Kosofsky, the Program Engineer for Lunar Orbiter, also attended.[8]

---

[6] Ibid.

[7] Ibid.

[8] Memorandum from Captain Lee R. Scherer to the Record, September 20, 1963.

Nicks expressed the belief that Headquarters and Langley had to maintain a well-defined, firm understanding on major policies to ensure the success of the whole undertaking. He sought from the beginning, through meetings such as this, to establish strong links of communication between the two groups in order to expose and resolve any problems quickly rather than allowing them the opportunity to grow into a major crisis for the program.

Thompson emphasized the importance of achieving an early understanding on all responsibilities by those in the program. There could be no room for inference; instead each member of the Lunar Orbiter Program had to recognize and agree upon an explicit basis for understanding what he was to do. The early establishment of a fixed point of reference from which future changes could be worked out was essential to the conduct of the program.

The September 11 meeting clarified the position of Headquarters and Langley. Each organization's representatives sounded out the others about delegation of authority and responsibilities. This approach was to be characteristic of relations between Langley and Headquarters throughout the program.[9]

---

[9] Thompson interview, January 29, 1970.

The Langley Source Evaluation Board

During September the Lunar Orbiter Project Office at Langley established the Source Evaluation Board (SEB) which it divided into several teams of experts who would analyze every contract proposal which they received. As an important part of the SEB, the Lunar Orbiter Project Office formed the Lunar Orbiter Proposal Scientist Panel to consider the scientific merits of each bidder's approach. The members of this reviewing group were Clinton E. Brown and Samuel Katzoff from Langley, Jack Lorell from the Jet Propulsion Laboratory, Norman Ness from the Goddard Space Flight Center, Bruce Murray from the California Institute of Technology, and Robert P. Bryson from NASA Headquarters.[10] They helped in the critical phase of proposal analysis, which began in October and lasted more than six weeks.

Of the score of possible aerospace companies which seemed to have the capability to carry out the objectives of a lunar orbiter program, five submitted contract proposals. To understand the significance of the spacecraft proposal which NASA finally chose, it will be useful briefly

---

[10] Memorandum from the Office of Lunar and Planetary Programs, NASA Headquarters, to Clifford Nelson, Project Director, Lunar Orbiter Office, Langley Research Center, October 22, 1963. See also: Agena Class Lunar Orbiter Photographic Project Plan for the Evaluation of Offerors' Proposals, Approved: Eugene C. Draley, Chairman, Source Evaluation Board, September 20, 1963.

to summarize the five choices which industry presented, remembering that NASA wanted a lunar orbiter which would require as little development of systems and as much use of off-the-shelf hardware as possible.

The Lunar Orbiter Proposals

The Hughes Aircraft Company, one of the five bidders, entered the competition with an impressive record. The Surveyor systems contractor for JPL, Hughes was no newcomer to the field of spacecraft design and fabrication. Its proposal centered on a spin-stabilized spacecraft. However, the Source Evaluation Board found in the Hughes approach several important weaknesses. First, while spin-stabilization greatly simplified the problem of attitude control, it placed disadvantages upon the photographic, power, and communications systems. Several inherent drawbacks in the photographic system, which would require extensive development before it could be incorporated into the spacecraft, compounded these disparities.[11]

The insufficiency of the power system to supply the necessary electricity to drive the other systems added a second negative aspect to the Hughes proposal. The SEB found that the design did not provide enough solar

---

[11] Memorandum for Lunar Orbiter Contract File, Subject: Debriefing of the Hughes Aircraft Company, Culver City, California, January 21, 1964, Langley Research Center, Hampton, Va.

cells to produce the required electrical energy and that if more were added Hughes would be forced to change the configuration of its spacecraft. In addition the proposal had given an incomplete description of the communications system, leaving out items which NASA had specified in the Request for Proposal document.

Finally, the Source Evaluation Board concluded that the solid-fuel retro-rocket for deboosting the spacecraft into lunar orbit was inadequate to alter the orbital parameters around the Moon. All of these factors, taken together, constituted too great an element of unreliability, and this plus the development problems outweighed the strong points of the spin-stabilization concept.

The only other proposal for a spin-stabilized lunar orbiter came from Thompson Ramo Wooldridge/Space Technology Laboratories of Redondo Beach, California. The TRW/STL orbiter concept used spin-stabilization to control the spacecraft's attitude during the mission. This meant that it had to make the other major systems compatible with spin-stabilization. While the attitude control problem was easily solved, it put severe restraints on the photographic system. It would have to employ fast shutter speeds and a high-speed film which would be very susceptible to solar radiation fogging.

The use of a liquid developer in the film processing

system also presented greater risks than would accompany other existing photographic systems. Moreover, due to the absolute necessity to maintain constant image-motion compensation, the quality of resolution of a single exposure might vary considerably from one side of the film to the other. The proposed format of a single photographic frame was too narrow, requiring the camera to make a large number of frames of any given area on the lunar surface.[12]

If the TRW/STL photo-system was judged impracticably elaborate, the proposed communications system simply failed to meet the requirements of the NASA RFP. Neither the communications nor the power systems were capable of performing their functions for the minimum thirty-day spacecraft life span. Because of spinning, the solar panels of the orbiter could not produce adequate quantities of power at any given time to recharge the spacecraft's battery. Moreover, the capacity of the battery was such that it could not have accepted a greater recharging rate than it already had, even if the energy producing area of the panels were enlarged. This amounted in the final analysis to a proposal with too many areas open to critical development

---

[12] Memorandum for Lunar Orbiter Contract File, Subject: Debriefing of the Space Technology Laboratories, Inc., Redondo Beach, California, January 22, 1964, Langley Research Center, Hampton, Va.

problems.

Ironically NASA had based its earlier decision to have a lightweight lunar orbiter on the STL systems research. STL had proven the feasibility of an Agena-class orbiter, but its concept of an orbiter proved to be less practicable than that of another bidder.

While Hughes and TRW/STL could claim experience in the increasingly complex realm of designing, building, and flying automated space probes, the Martin Company, which offered a third approach, had no such advantage in this respect. However, it presented a very satisfactory proposal from the standpoint of technical feasibility. Unlike the first two firms, Martin designed its orbiter to employ three-axis stabilization to serve as the attitude control system for a platform from which a very well-designed photographic system could take pictures of the Moon without having to compensate for rate of spin.

Although it had a limited capability to perform high-quality convergent stereo photography, its film processing, readout, and communications systems appeared to be highly capable of transmitting data to Earth in a very short time. This aspect of the Martin proposal greatly pleased the SEB evaluators at Langley. On the other hand, the Martin orbiter lacked redundant systems which would ensure greater reliability in spacecraft performance, and the proposed solar panels seemed to the Source Evaluation Board somewhat fragile for the task of supplying energy to the spacecraft.[13]

Martin's proposal showed its most serious weaknesses in the areas of launch and flight operations and in the use of the tracking and data-acquisition facilities. The proposal stressed launch operation procedures over flight operations, and the description of both was ambiguous. Moreover, Martin had failed to include an integrated plan of the functions and responsibilities of NASA, Martin, the Deep Space Instrumentation Facility, and the Space Flight Operations Facility and their personnel. Finally, because of limited experience in spacecraft design and fabrication, Martin would necessarily have to rely upon subcontractors, and this could present NASA with major difficulties in the event that relations between Martin and its subcontractors became disturbed. This, according to the SEB, made the Martin proposal the least practicable from the standpoint of program management.[14]

The two remaining bidders -- the Lockheed Missiles and Space Company and the Boeing Company -- presented the

---

[13] Memorandum for Lunar Orbiter Contract File, Subject: Debriefing of the Martin Company, January 21, 1964, Langley Research Center, Hampton, Va.

[14] Ibid.

Source Evaluation Board with an interesting challenge. The former had long years of experience in designing and building the Agena system for the U.S. Air Force. Indeed, its Agena had served as a photographic platform in Earth orbit. The rocket and the photographic systems were well mated, making a very efficient spacecraft for work in orbit around the Earth. Lockheed proposed to convert this to an orbiter for lunar photography. It would consist of the Agena with integrated photographic, power, communications, and attitude control systems. Lockheed stressed that the Agena had been proved in space and would require only minor modifications, thus making it unnecessary for NASA to buy a new, expensive, and untested spacecraft.[15]

The Boeing Company, on the other hand, could not make such an offer, since it had never managed a major NASA space flight program. Aircraft manufacture was Boeing's big business, but competition in the aerospace industry motivated the Seattle-based firm to turn toward space projects and to invest in new capital equipment in order to meet and excel in the increasingly competitive world of rocket research and space exploration. Indeed as part

---

[15] Memorandum for Lunar Orbiter Contract File, Subject: Debriefing of the Lockheed Missiles and Space Company, Sunnyvale, California, January 21, 1964, Langley Research Center, Hampton, Va.

of the USAF Project Dynasoar, Boeing had constructed its new Kent Facility for testing spacecraft components under simulated space environmental conditions. This capability would enable Boeing to conduct its own testing without costly delays caused by the necessity to send equipment elsewhere to be tested. (Project Dynasoar was canceled about the time NASA became seriously involved in a new lightweight lunar orbiter.)

The Source Evaluation Board saw the facility with which Lockheed's proposal might be implemented and realized that Boeing did not have as much experience in spacecraft design and fabrication. But the Lockheed proposal had some serious flaws which outweighed the attractive possibility that NASA might obtain a ready-made orbiter.

First, the existing Agena system was designed for Earth orbit, and it had proved its ability to perform there very well. But sending a spacecraft some 385 kilometers into space and putting it into orbit around the Moon was an entirely different undertaking, and the configuration of the Lockheed orbiter presented special problems related to this. Any lunar orbiter would be useless if it could not orbit the Moon as NASA scientists and engineers desired it to do. Moreover, any orbiter would be a waste of money if it could not perform the desired photography in the most efficient, reliable way possible with existing technology. The SEB believed that the use of any incompatible hardware for such critical work would impinge upon mission assurance.

This being the case, the Source Evaluation Board found the concept of sending a modified Agena rocket to do lunar orbital photography too impracticable, because the Lockheed orbiter presented the extreme difficulty of deboosting the heavy deadweight Agena into a lunar orbit. Once deboosting was accomplished, the spacecraft's orbit would create severe restraints on photography. NASA would have to go to unnecessary trouble to obtain vital photographic data of the lunar surface, and this fact made the Lockheed proposal much less attractive.[16]

Yet the SEB found the Lockheed photo system to be almost ideally suited to its task. It was a space-proven package with the capability of performing high-quality stereographic photography. However, the proposed processing and readout systems would require more development before Lockheed could use them in an orbiter, and this meant extra time and funds to accomplish basic development work. Even if this were surmountable, the necessity to carry the heavy deadweight of the burned-out Agena to the Moon still remained the major negative factor of the Lockheed Orbiter. It would require extra fuel to control the useless

---

[16] Ibid.

bulk in lunar orbit. Hardly any of the Agena's weight would be directly involved in vital mission activity, and yet its presence would definitely affect orbital parameters and spacecraft velocity to the extent of reducing the versatility of the orbiter as a photographic platform. These features made the Lockheed approach less acceptable than that of the final bidder.

The Boeing Lunar Orbiter Proposal

The Source Evaluation Board turned to the proposal of the Boeing Company of Seattle, Washington. Boeing presented an orbiter concept which used three-axis stabilization with a spacecraft weighing only 360 kilograms. The design employed much space-tested, off-the-shelf hardware. For example, Boeing would have a photographic system fabricated by Eastman Kodak, the contractor for the Agena photo system already in use by the U.S. Air Force. Film processing on board the orbiter would be handled by the Kodak Bimat process which had been perfected in 1961. The Boeing orbiter would use the same Canopus sensor for acquiring the star Canopus as an attitude reference as the Mariner C spacecraft had used. The 100-pound-thrust Marquardt rocket engine which was being developed for the Apollo Program would be used for deboosting the spacecraft into lunar orbit. Four large solar panels would generate power for the spacecraft, and these would be backed up by nickel cadmium batteries which would supply power at the times when the orbiter would be out of sight of the Sun. The whole system would generate 266 watts of electrical output to power the spacecraft's components.[17]

Boeing's proposed photographic system pleased the Source Evaluation Board because it offered greater flexibility than those submitted by the other four bidders. It would be a scaled-down version of the Eastman Kodak system used by USAF, and, unlike the others, it featured a camera with two lenses which could take pictures simultaneously -- one using a high-resolution, the other a medium-resolution mode. On a single mission the Boeing orbiter could photograph a greater area of the lunar surface and also obtain more detailed photographic data than any other proposed system. Moreover, if loss of the use of one lens occurred, the whole photographic mission would not be ruined.

The photographic system would be capable of providing pictures of areas up to 8,000 square kilometers in the high-resolution mode -- four times the size of area called for in the NASA Request for Proposals. Moreover, the photographic payload would use the very suitable, highly perfected Kodak

---

[17] OSSA Review -- Lunar Orbiter Status Report, January 23, 1964, p. 2.

Bimat process to develop and fix the film on board the spacecraft. It is, therefore, important to the understanding of the Boeing lunar orbiter concept to survey briefly the photographic system and the Bimat process in order to recognize the greater degree of flexibility which these two integrated subsystems offered NASA.

The Eastman Kodak Photographic System

The basic system which Eastman Kodak would provide Boeing had been in existence since mid-1960, when Kodak had developed it for military applications. For Boeing's use it had been reduced in size and weight to fit within the Agena weight restrictions. The mechanics of the system were as follows: Film from a supply reel passed through a focal plane optical imaging system, and controlled exposures were made. Once past the shutter, the film underwent a semi-dry chemical developing process and then entered a storage chamber. From here it could be extracted upon command from the ground for scanning by a flying-spot scanner and then passed on to a take-up reel.

The line-scanning device consisted of a cathode-ray tube with a rotating anode having a high-intensity spot of light. The scanner optics of the moving lens system reduced by 22 times this point of light, focused it on the film transparencies and scanned them. A photomultiplier then converted the light passing from the scanner through the film into an electrical signal whose strength would vary with the density of the emulsion layer of the film. This signal would then be transmitted to a receiving station on Earth and reconstructed. The Eastman Kodak Company would upgrade the system for the demands of the Boeing orbiter and its mission.

A significant part of the improvement in the system was the introduction of the Kodak Bimat process, which eliminated the necessity to use "wet" chemicals on the film. Instead, a film-like processing material was briefly laminated to the exposed film to develop and fix the negative image and, if the need existed, to produce a positive image. In the case of the Boeing orbiter this second step was not used, and only negatives were made.[18] Once the film had been developed and fixed, the Bimat material separated from the film and wound onto a storage spool.

Kodak's "dry" process offered the photographic system of the Boeing orbiter very positive advantages over those of the other bidders. Besides eliminating the need for liquids and their storage containers, Bimat did away with the necessity of an extra fixing step while producing

---

[18] Raife G. Tarkington, "The Kodak Bimat Process," Photogrammetric Engineering, Vol. XXXI, No. 1 (January 1965), p. 126.

photographic negatives having normal, high-quality physical, sensitometric, and image characteristics. This greatly simplified the problems involved in materials-handling while making the whole process fully automatic. Moreover, every part of the film enjoyed fresh-processing chemistry, which made the resulting negatives more consistent and uniform. Bimat would not leave any crystalline deposit on the film after separation, and lamination of the two materials would not result in any damage to the emulsion layer. In addition, the position of the equipment would not affect processing of the film, a factor which made the Bimat process ideally suited to work in a space environment.[19]

The Boeing-Eastman Kodak photographic system was not the only strength of the proposal. Boeing also demonstrated a very real understanding of the relationship of the various program phases to one another as detailed in the Request for Proposals. It clearly expressed its willingness to cooperate with NASA and to keep a nucleus of full-time personnel managing key areas of the program from the beginning to the conclusion of operations. Proven technical competency, flexibility and imagination, sound planning and organizational management, wide use of space-tested hardware in the spacecraft design, reliable test facilities,

---

[19] *Ibid.*

and the absence of any major development tasks or the need to rely on many subcontractors made the Boeing Company's lunar orbiter proposal the most realistic, manageable, and potentially successful of the five. The NASA-Langley Source Evaluation Board overwhelmingly graded Boeing's proposal as the most likely to fulfill the objectives of the Lunar Orbiter Program and to cost the least per photograph returned to Earth.

Selecting the Lunar Orbiter Contractor

The final decision on which of the five proposals to choose rested with NASA Associate Administrator Robert C. Seamans, Jr. The Langley SEB recommended that NASA select Boeing. Thompson passed his center's recommendation on to Seamans. Yet Seamans had to be convinced not only that the proposal's technical approach was the best, but also that its management arrangements and estimated costs were better than those of the other bidders. Boeing seemed to meet two of the three criteria, but its cost figure was substantially higher than that of the next nearest bidder -- Hughes.

Seamans had to find an absolute justification for selecting the highest priced bid in order to defend the choice before Congress if called on to do so. That absolute factor turned out to be a technical detail of major significance for the success of the Lunar Orbiter Program.

Dr. Trutz Foelsche, a Langley scientist working in the field of solar radiation hazards, had been conducting experiments whose results demonstrated that even small doses of radiation from solar particle events were "of major importance for such sensitive devices as, e.g., photo-emulsions or ordinary photographic films, which are an important tool in some space missions. This is especially true for instrumented probes, when the vehicle itself generally provides shielding only on the order of 1 g/cm^2 or less from a large solid angle."[20] Foelsche's data, based upon the largest solar event groups of the 1954-1964 sunspot cycle, showed that high-speed films did not receive sufficient protection even when shielding around the film was increased up to 10 grams per square centimeter. (See chart on the following page for Foelsche's data.)[21]

Foelsche presented his findings to Dr. Thompson and the Source Evaluation Board before the final selection of the Lunar Orbiter contractor. The Langley SEB made a presentation to Dr. Seamans and senior OSS staff members at NASA Headquarters in November 1963. Following this, Seamans met with NASA Administrator James E. Webb and NASA

---

[20] Dr. Trutz Foelsche, "Remarks on Doses Outside the Magnetosphere, and on Effects Especially on Surfaces and Photographic Films," paper presented at the Meeting to Discuss Charged Particle Effects, NASA, Office of Advanced Research and Technology, March 19-20, 1964, Washington, D.C., p. 8.

[21] Ibid.

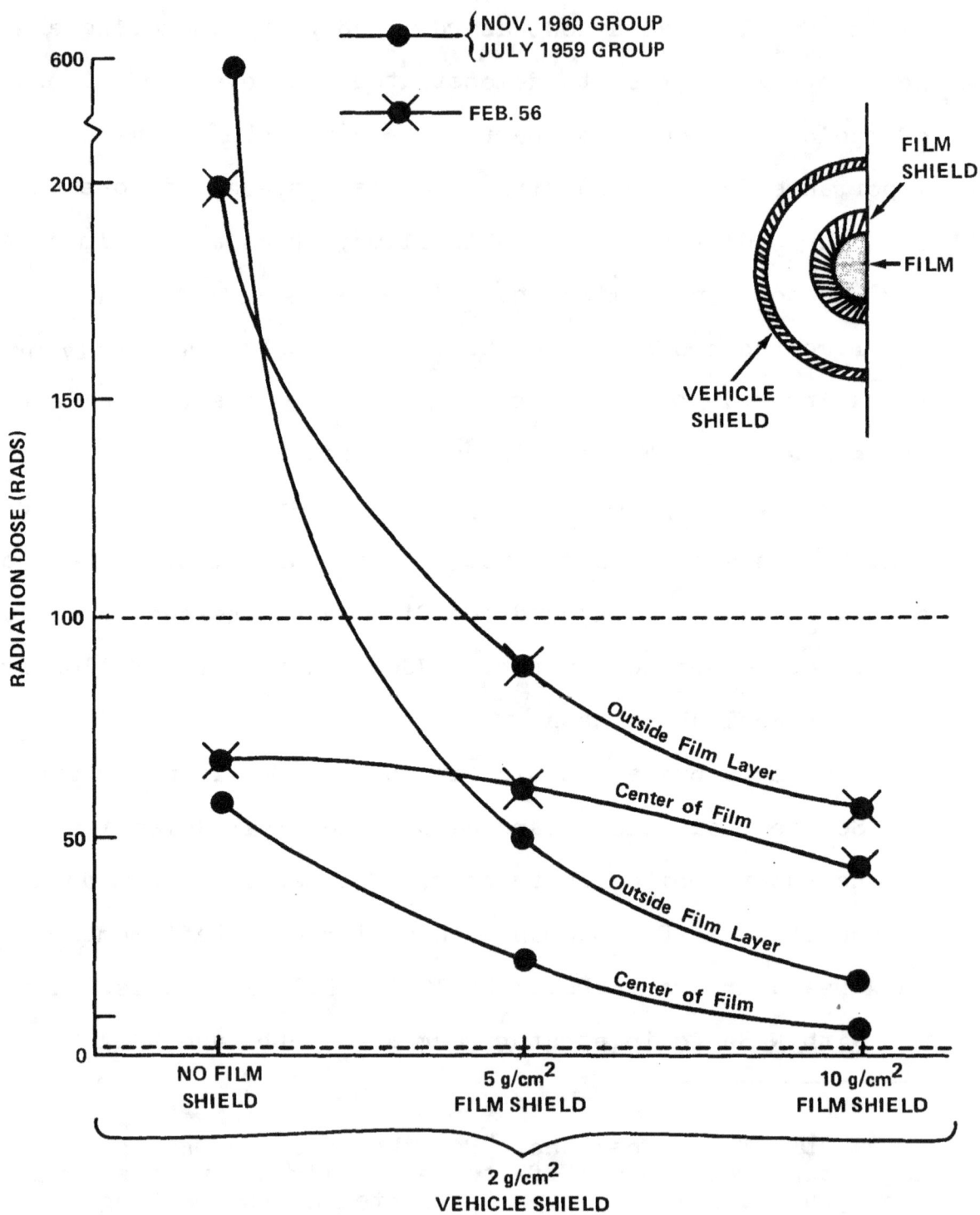

Upper Limits of Doses on Surface and in Center of Film Sphere Surrounded by Additional Shields for Largest Solar Event Groups of the Sunspot Cycle 1954-1964.

RADIATION DOSES

Deputy Administrator Dr. Hugh L. Dryden. The three conferred and agreed that Seamans would meet separately with representatives from each of the five companies in order to develop a better understanding of each proposal's technical aspects.[22]

Dr. Seamans arranged for each bidder to brief him and Earl D. Hilburn, NASA Deputy Associate Administrator for Industry Affairs, together with several members of the Langley Source Evaluation Board. The briefings took place in Washington over a week-long period. The data on radiation hazards to film enabled Seamans to question each bidder from a position of strength about the problem of film damage in their systems due to a possible solar particle event during the thirty-day mission which an orbiter would have to carry out.

The two bidders who had proposed spin-stabilized spacecraft necessarily had to rely on high-speed film and fast shutter speeds to compensate for image-motion. Two other bidders also had their photographic systems designed to employ high-speed films. When asked directly what would happen in the event of a solar flare, they had to

---

[22] Letter from Dr. Robert C. Seamans, Jr., to Dr. Eugene M. Emme, NASA Historian, Washington, D.C., Comments on "Lunar Orbiter: A Preliminary History," Comment Edition (HHN-71), November 25, 1969.

admit that their film would incur significant damage.

Only the Boeing-Eastman Kodak system was designed to use a very low speed, insensitive film (ASA @ 1.6) which, with minimal shielding, would not be endangered by sudden discharges of high-energy radiation from the Sun or during transit through the Van Allen belts.

Seamans concluded with confidence that the Boeing proposal definitely offered NASA advantages and safeguards which the other proposals did not. He concurred with Langley's recommendation that NASA choose Boeing as the contractor, and this decision opened the next phase of the program.

CHAPTER IV

NASA AND BOEING NEGOTIATE A CONTRACT

Early Boeing Preparations

The Boeing Company of Seattle, Washington, had been among the bidders for the Apollo Program's Lunar Excursion Module (LEM, later called Lunar Module, or LM) and had lost the competition to the Grumman Aircraft Corporation in the spring of 1963. Boeing's research studies for the LEM proposal enabled a team led by Thomas Yamauchi in the Aerospace Group to develop data for lunar orbital missions. The technical expertise which Boeing had assembled during the work on the LEM proposal subsequently became available for new work on an unmanned lunar orbiter. Boeing began to develop a proposal for a lunar orbiter spacecraft during the summer of 1963, utilizing the earlier research work it had done for its LEM proposal.[1]

When Boeing presented its proposal to the NASA-Langley Source Evaluation Board it had developed and analyzed a spacecraft system whose capabilities matched or exceeded the requirements of the RFP. The Boeing proposal appeared so complete in its coverage of the technical problems of

---

[1] Recorded interview with Thomas R. Costello, Aerospace Group, The Boeing Company, Washington, D.C., July 9, 1970.

creating a lunar orbiter that if the members of the SEB were to find any part of it questionable they would be forced to challenge the original assumptions upon which the Request for Proposals had been based.

Among other key system problems, Boeing Company had even analyzed the possible danger to the camera film from radiation. From its analysis, Boeing developed data showing that high-speed films were subject to degradation and fogging if they were not properly shielded from solar-flare-particle events. When Boeing convinced the Eastman Kodak Company to build the photographic system for its lunar orbiter, the data on radiation fogging of film enabled both to select a low-speed, insensitive film which would, nevertheless, perform the photographic tasks outlined in the RFP.

The Boeing proposal won the NASA-Langley recommendation for acceptance, and on December 20, 1963, NASA Administrator James E. Webb announced the selection of Boeing to build Lunar Orbiter.[2]

The Boeing Company had already established its Lunar Orbiter Program Office in June 1963 under the direction of Robert J. Helberg. Between June and December Helberg had handled the complete management responsibilities for

---

[2] OSSA Review -- Lunar Orbiter Program Status Report, January 23, 1964.

the 220-man Lunar Orbiter Team. He organized a tightly knit project group and directed its members in the preparatory activities of the Lunar Orbiter proposal. These included research, technical design, test program analytical studies, the reliability program, manufacturing, quality assurance, contract administration, finance, facilities, and program controls. Helberg was a very capable administrator with an engineering background and, since 1958, experience in the Bomarc Program.[3]

Boeing selected George H. Hage to assist Helberg as the Chief Engineer of the Lunar Orbiter Program. Hage had been a member of the Lunar Excursion Module Engineering Team, and early in 1963 he had also taken charge of new business in the area of lunar reconnaissance. He directed studies and preliminary designing in the development and definition of an unmanned lunar orbiting satellite designed to obtain high-resolution photographic data of the Moon's surface. Following this Hage had handled Boeing's technical activities during its proposal effort on the Agena-class Lunar Orbiter Project.[4]

Carl A. Krafft was assigned to be the Lunar Orbiter Program Business Manager. Coming from the Bomarc Branch, he had experience in operations planning, costs and expenditures control, performance evaluation, administration,

---

[3] Boeing Company biographical note on Robert J. Helberg.
[4] Boeing Company biographical note on George H. Hage.

and progress reporting. While with the Bomarc Branch he had directed the use of the PERT/Time and PERT/Cost and Line-of-Balance control techniques. (PERT stands for Performance Evaluation Reporting Technique.) Krafft had gained extensive experience in contract negotiation, in accounting for contract execution, and in the preparation of work statements and contract proposals.[5]

Two events augured well for the establishment of the Lunar Orbiter Program at Boeing. First, the building housing the Bomarc Program became available to Helberg, and he moved his organization in under one roof. At the peak of the program Boeing had 1,700 to 1,800 people working on Lunar Orbiter. The large, isolated facility accommodating Helberg's organization made communications between various members of the Lunar Orbiter Program more open and nearly instantaneous.

Secondly, the U.S. Air Force canceled Project Dynasoar in the spring of 1963, releasing a number of highly qualified resident USAF personnel members to support Boeing's new NASA undertaking. Some of the USAF people had been engaged at Boeing on the X-20 Project, and they also became available for work on Lunar Orbiter. The Air Force personnel worked in two areas: engineering monitoring

---

[5] Boeing Company biographical note on Carl A. Krafft.

and quality control. In both they assisted Boeing with their specific technical expertise. This assistance saved manpower at Langley.

NASA Preparations for Contract Negotiations

On November 1, 1963, Dr. Homer E. Newell announced the details of an organizational change which merged the Office of Space Sciences and the Office of Applications to form the new Office of Space Science and Applications (OSSA). This new organization became the Headquarters base for the Lunar Orbiter Program. The Office of Lunar and Planetary Programs, directed by Oran W. Nicks, was a division of OSSA.[6]

After the Christmas holidays, preparations for the NASA-Boeing contract talks got under way on January 6. The Office of Space Science and Applications sent Headquarters representatives to Boeing together with Langley contracting officers. The conference there resulted in an agreement on basic task areas which NASA and Boeing would work out before signing a contract. They also drew up a tentative schedule of activities for the following sixty days.

Following the Boeing meeting Langley officials met

---

[6] Memorandum from Associate Administrator for Space Science and Applications to Division Directors, Office of Space Science and Applications, November 1, 1963.

with officials at the Jet Propulsion Laboratory to establish preliminary agreements on how Langley might best benefit from JPL assistance. JPL people pointed out at this time that problems involving trajectory design for Lunar Orbiter would have to be handled by Langley and Boeing. Trajectory design, with its known strong correlation to the internal design of the spacecraft, could not easily be done by JPL without JPL becoming involved in spacecraft design. This kind of involvement would place a severe burden on the manpower situation at JPL and would constitute the probable germ of interlaboratory friction.

JPL officials defined the facility limits in tracking time and the probable ways in which the Deep Space Net (DSN) could best serve Lunar Orbiter. The tracking and data-acquisition facilities at JPL and the DSN were serving the needs of Ranger, Mariner, Surveyor, and Pioneer and Centaur during the period in which the Lunar Orbiter Program was establishing itself. JPL made an additional commitment to serve the needs of Lunar Orbiter when the time came to fly.[7]

Following the West Coast preparations, NASA-Langley

---

[7] Letter from Dr. Eberhardt Rechtin, Director, Advanced Research Projects Agency, Washington, D.C., to Dr. Eugene M. Emme, NASA Historian, November 18, 1969, with comments on manuscript "Lunar Orbiter: A Preliminary History" (HHN-71).

representatives met with officials of the Lewis Research Center and the Lockheed Missiles and Space Company, the prime contractor to Lewis for the Agena launch vehicle. At this time an intercenter agreement was established to cover the Agena-Lunar Orbiter interface. Subsequently the Lunar Orbiter Program Office in Washington conducted an information meeting to acquaint representatives of the various government mapping agencies with the Lunar Orbiter spacecraft design and the NASA mapping requirements as they existed at the time. By late January Boeing officials at Langley completed the preliminary tasks required for actual contract negotiations and gave a detailed presentation of all elements of their proposal with tentative cost estimates and funding requirements.[8]

Lunar Orbiter planning accelerated during February when NASA officials met again with the Air Force personnel stationed at Boeing to discuss the role which they would play in the Lunar Orbiter Program. Following this meeting the Office of Space Science and Applications drafted a document defining the USAF support activity and sent it to Langley and the Air Force for approval.

The Lunar Orbiter Project Office at Langley desired to make as much use of Air Force technical support at

---

[8] OSSA Review -- Lunar Orbiter Program Status Report, January 23, 1964.

Boeing as possible, especially since the Air Force had extensive experience with the Eastman Kodak camera system. In addition Boeing representatives met at Langley with officials from Lewis to discuss the problems of integrating the Agena and the spacecraft systems and to distribute the responsibilities involved in this task. Boeing and NASA officials agreed that Lewis would handle the shroud which would enclose the Lunar Orbiter atop the Atlas-Agena launch vehicle. Eventually Lewis issued an RFP for the shroud. It awarded the contract to Boeing and supervised production of the shroud. Once Boeing realized that Lockheed, manufacturer of the Agena, would not be able to handle the shroud, Boeing decided to take responsibility for its design and manufacture. Boeing wanted to see that the shroud and the spacecraft were absolutely compatible.

In addition to making the shroud Boeing would take care of the adapter and separation systems, which would integrate the spacecraft-shroud combination with the Agena and separate them at the proper time in space.

Other Boeing officials continued to work out cost estimates with Langley contracting officers, and Langley finished drafting an integrated work statement toward the end of February. These preparations enabled NASA/Langley to begin detailed contract negotiations with Boeing, and on March 2 the talks commenced.[9]

Congressional Criticism of Contractor Choice

While the Office of Space Science and Applications, the Langley Research Center, and the Boeing Company proceeded to work out the fine points of the Lunar Orbiter contract, some congressional criticism over NASA's choice of contractors rumbled down from Capitol Hill to NASA Headquarters. According to Aviation Week & Space Technology, NASA had decided to choose the Boeing proposal "because it offered the greatest assurance of mission success," and although the Seattle firm's price tag was seemingly the most expensive (approximately $60 million) "the firm won the contract because of the high reliability factor in spacecraft design approach."[10]

As satisfying as this may have been to NASA and Boeing, it struck a dissonant chord with Congressman Earl Wilson of Indiana. Wilson questioned NASA's selection of Boeing's more expensive bid over that of the Hughes Aircraft Company, which would have cost supposedly half as much. The Space Science Subcommittee of the House Committee on Science and Astronautics, chaired by Congress-

---

[9] Status of Lunar Orbiter Program for possible use in OSSA Review, February 24, 1964.

[10] "Boeing to Build Lunar Orbiter," Aviation Week & Space Technology, Vol. 79, No. 27 (December 30, 1963), p. 22.

man Joseph Karth of Minnesota, joined Wilson and questioned NASA spokesmen extensively about their choice of Boeing. Despite their criticism NASA succeeded in convincing the Congressmen that "Boeing's proposal was selected because of its three-axis system rather than the spin-stabilized system suggested by Hughes."[11]

Although one approach was not necessarily better than the other, the three-axis system greatly reduced the technical difficulties involved in the photographic system. Moreover, the Boeing proposal had a far superior technical approach to obtaining the necessary photographic data and a greater inherent likelihood that it would reliably do just that. This had been the determining factor in the evaluations of the five bidders' proposals. Langley evaluators had employed the philosophy that the price of a proposal was secondary to the quality of the technical design and the management program which the bidder offered. In both respects the Boeing bid had been judged superior.

No Duplication of Effort

Having vaulted the congressional hurdle, OSSA turned next to examine suggestions within NASA of the possible

---

[11] "NASA Explains Choice of Boeing Over Hughes in Lunar Orbiter Award," *Missiles and Rockets*, Vol. 14, No. 10 (March 9, 1964), p. 13.

duplication of work and development in the Lunar Orbiter Program. Earl D. Hilburn, Deputy Associate Administrator for Industry Affairs, notified Edgar M. Cortright in OSSA early in March that his office was concerned about the apparent intention of the Lunar Orbiter Program Office to allow Boeing to develop a new attitude control system despite the fact that NASA had already invested $10 million in research and development for such systems for the Ranger and Mariner spacecraft. Hilburn pointed to the possibility that Boeing might desire to use the Lunar Orbiter contract as a means to justify building up a new technological capatility. Hilburn requested that Cortright scrutinize any such situation in contract negotiations with Boeing and establish a reason for any seeming duplication of effort.[12]

Cortright responded to Hilburn quickly with a lengthy description of the NASA-Boeing negotiations as they had developed through March. The Lunar Orbiter Program, he stressed, was attempting to make the maximum use of flight-proven hardware. This meant that Boeing would serve as the prime systems integrator because it alone retained the

---

[12] Memorandum from Earl D. Hilburn, Deputy Associate Administrator for Industry Affairs, to Edgar M. Cortright, Deputy Associate Administrator for Space Science and Applications, March 19, 1964.

responsibility for the Lunar Orbiter spacecraft structure and attitude control system. Boeing and NASA would spend more than 50% of the contract funds on hardware which Eastman Kodak and RCA would supply.

Contrary to Hilburn's major worry, the Boeing Company had a well-developed electronics capability gained through its experience as contractor for the Bomarc, Dynasoar, and Minuteman systems, and despite this NASA negotiators had encouraged Boeing to look for companies with greater competency in guidance systems: Northrop, Philco, General Electric, and Bendix, for example. Moreover, during the final phase of the Ranger Program when a fifth block of spacecraft had been under consideration, Northrop had been prime contractor. When the Block V Rangers were canceled in December, 1963, Northrop had been assigned to conduct a technology transfer study. This study had proved very useful to NASA and Boeing.[13]

Cortright stressed that the Lunar Orbiter Program Office and the Boeing Company were basing contract talks on the axiom that they use as much off-the-shelf hardware

---

[13] On March 8, 1963, NASA had announced the selection of the Northrop Corporation for industrial support on Ranger Blocks III and IV and as contractor for producing Ranger Block V spacecraft (see Aviation Week, March 18, 1963). On December 13, 1963, NASA Headquarters directed JPL to terminate all activities with the Ranger Block V (see NASA Astronautics and Aeronautics, 1963, p. 477). Following this, Northrop began a technology transfer study (see Northrop Space Laboratories, Technology Utilization Review and Analysis, Final Report, Vol. II, NSL 64-192, September 1964).

as possible.[14] He stressed that because the attitude control system of the Lunar Orbiter spacecraft would have to fulfill many more demands than that of a Ranger or a Mariner deep space probe, and because the system was so interrelated to all other spacecraft systems, the Office of Space Science and Applications had decided that the prime contractor, Boeing, should take the full responsibility for the attitude control system and its integration with all other systems. However, NASA and Boeing had reached agreement that the latter would use at least the following items of hardware in building the attitude control system:

1. Inertial Reference Unit -- to be purchased from Kearfott, previously used on Mariner C.

2. Sun Sensor -- to be purchased from Bendix, previously flight qualified.

3. Canopus Sensor -- identical with one on board Mariner C; JPL fabricating this item. Boeing would request proposals from seven contractors, including Northrop, using JPL specifications.

4. Reaction Control System (thrusters, squibs, filters, regulators, etc.) -- to be purchased from various companies. Boeing to construct the nitrogen tanks.

5. Flight Programmer -- because of the complexity and critical importance of this unit, Boeing would retain full responsibility but would purchase items for its construction from various companies as it

---

[14] Memorandum from Edgar M. Cortright to Earl D. Hilburn, April 8, 1964.

deemed fit.[15]

The brain of the spacecraft would be the Flight Programmer, an electronic wizard approximately the size of a shoe box, and its performance could determine the success or failure of any mission to the Moon. Because of the crucial role of the Flight Programmer, its configuration significantly influenced the design of the rest of the Lunar Orbiter's systems. (See Chapter VI for a description of the Flight Programmer.) The completion of the Programmer would have to await the integration of the spacecraft's other components and subsystems so that it could be placed in the spacecraft as the nerve center linking all of the parts together in an electronic organism.

Langley and the Office of Space Science and Applications believed that Boeing had to retain the complete responsibility for the Programmer, the attitude control system, and their integration. Boeing also would conduct any necessary analyses, engineering, and computer studies of this system in order to have the working flexibility to cope with unforeseen problems and unexpected changes.[16] This arrangement in no way meant that Boeing would under-

---

[15] Ibid., p. 2.
[16] Ibid.

take the completely new design and fabrication of a unique attitude control system. On the contrary, the record demonstrated convincingly that the contractor was attempting to use as many off-the-shelf and flight-proven items of hardware as possible and that it was utilizing experience gained in earlier NASA programs.

NASA Solely Responsible for Photographic Data

A more difficult problem impinging upon contract negotiations was the working relationship which Boeing and NASA were going to establish with the two major subcontractors: RCA and Eastman Kodak. Eastman Kodak's photographic system would be the heart of the Lunar Orbiter, and this meant that Eastman Kodak would play a major role in the success of the program. However, NASA-Langley and Boeing had to define and limit the extent of this firm's participation in the Lunar Orbiter Program.

One reason for this became apparent when Boeing suggested that the Lunar Orbiter Program use the Eastman Kodak facilities for reconstituting and processing photographic data from the spacecraft. Boeing considered this to be advantageous because of the presence of the NASA-owned Ground Reassembly Printer at the EK plant in Rochester New York.[17] Lt. Col. Clifton E. James, Assistant for

---

[17] Memorandum from Dr. Homer E. Newell, Associate Administrator for Space Science and Applications, to Dr. Robert Seamans, Associate Administrator of NASA, March 19, 1964.

Photography, USAF Office of Space Systems, raised the first sign of disapproval of the Boeing idea in a memorandum to Brockway McMillan, the Under Secretary of the Air Force, in February. James stressed that "the achievement of large scale lunar photography will most certainly create wide public interest which can be compared with the acclaim accorded to Sputnik I and the first manned orbital flight."[18]

Because of the great potential impact of such an event and because it would be sustained not by one but by five photographic missions, James felt that United States space exploration would best profit if the National Aeronautics and Space Administration managed every facet of the processing, handling, and distribution of all photographic and other data transmitted to Earth by the spacecraft. James stressed that "the selection of a contractor's facility for establishing the Lunar Photographic Production Laboratory will not only detract from the potential prestige of this program, but it will also result in management problems...."[19]

In NASA Seamans read the James memorandum and sent it on to Homer E. Newell in OSSA for review. After evalua-

---

[18] Memorandum from Lt. Col. Clifton E. James, USAF Office of Space Systems, to the Under Secretary of the Air Force, February 26, 1964, p. 1.

[19] Ibid., p. 3.

ting the criticisms which James had raised, Newell's office resolved that, although "the consequences of performing this work at Eastman Kodak are uncertain, the possible disadvantages appear to outweigh the advantages."[20] Newell felt that Eastman Kodak, with its reputation for extremely precise, high-quality work but also strong security consciousness, might hinder the accessibility of interested parties to the lunar photographic data. Therefore, his office recommended that NASA conduct the processing of Lunar Orbiter photographic data, most likely at Langley, using technicians from EK in the initial stages of data reduction. All of this work would be done under NASA auspices and management. Boeing would have to accept NASA's position on this matter as final.

Langley-JPL Working Relations

Langley began to work with the Jet Propulsion Laboratory in the establishment of the formal support activity which the Lunar Orbiter Program would require in order to fly the five authorized missions. Members of the Lunar Orbiter Project Office at the Langley center met with JPL officials during the spring of 1964. The vital service which the JPL-managed Deep Space Net, consisting of the Deep Space Instrumentation Facility (DSIF) and the Space

---

[20] Memorandum, Newell to Seamans, March 19, 1964.

Flight Operations Facility (SFOF), would provide Langley was stated as "the acquisition, transmission, processing, display, and control of spacecraft tracking and communications information necessary to the support of flight project mission requirements. These project requirements include navigation, scientific measurements, photography, spacecraft and mission control, and spacecraft performance monitoring."[21]

Eventually the JPL DSN support effort for Lunar Orbiter approached the level of between 500 and 1,000 man-years of work. At the same time the tracking and data-acquisition facilities also served the Ranger, Mariner, and Surveyor programs. At first Langley experienced some difficulties in defining precisely what tasks JPL could perform for the program, but this was no fault of JPL. On the contrary, JPL, facing manpower shortages and a scarcity of computer time, managed to meet the needs of the Lunar Orbiter Program without causing any schedule slippages or launch delays.[22]

One of the key problems in establishing a coordinated working relationship between Langley and JPL was the defi-

---

[21] J. R. Hall (ed.), <u>TDS Final Report</u>, Tracking and Data System Report Series for <u>Lunar Orbiter Project</u>, Vol. I, <u>Support Summary</u> (608-15), Jet Propulsion Laboratory, September 1, 1969, p. 1-1.

[22] Letter, Rechtin to Emme, November 18, 1969.

nition of the extent to which JPL should become involved in analytical work for Orbiter, involving such areas as trajectory design. Langley requested JPL to make a definitive study of the Lunar Orbiter tracking data requirements to parallel a similar one which Boeing was conducting. At the Lunar Orbiter Mission and Trajectory Analysis Meeting on April 15, JPL representatives suggested to Langley officials that Boeing send one or more men to undergo a familiarization and orientation period at the DSN facilities so that Boeing might know exactly what the facilities offered. Following this Boeing could erect its own computer facility to simulate the Space Flight Operations Facility, accomplish its own programming, and check out and integrate this set-up with that of JPL at SFOF.

The problem which Langley and Boeing had to work around was the shortage of computer time at the JPL facilities due, in part, to the needs of Surveyor. The familiarization and orientation period would involve approximately 20 man-years of work. More important, however, for JPL was the recognition that any direct and intimate involvement in trajectory design and related analyses would demand that JPL also become involved in spacecraft design, because much of the planning of software and trajectory design depended upon the design of the spacecraft's communications system. JPL, understandably, was not in a position

93

to commit manpower and computer time to such work for Langley, and it made this clear in a memorandum to Floyd L. Thompson on April 2, 1964. Following the April 15 Trajectory Analysis Meeting Thompson notified Newell at NASA Headquarters of the JPL position.[23] The JPL suggestion to educate Boeing men at its DSN facilities proved acceptable to Boeing and Langley.

In addition to meetings with JPL officials, Lunar Orbiter Project officials from Langley spent two days at the beginning of April with representatives from Boeing and OSSA at the Kennedy Space Center inspecting the facilities for Lunar Orbiter. They also briefed personnel there on the Orbiter requirements which KSC would have to meet. Scherer noted that the program needed new hangar facilities at Cape Kennedy if it wanted to avoid an undue burden on existing space.[24]

With most of the anticipated problems resolved, the Langley Research Center and the Boeing Company signed the Lunar Orbiter contract on April 16 and sent it to NASA Headquarters for final review. The total period of contract negotiations had been remarkably short and intense.

---

[23] Ref.: (a) Memorandum to NASA Code S, Attention: Homer E. Newell, from Langley Director, Subject: Request for Additional Support for Lunar Orbiter from JPL, dated April 2, 1964, dictated by Crabill (LRC), April 20, 1964.

[24] OSSA Review, Lunar Orbiter Status Report, May 5, 1964.

NASA and Boeing worked out an excellent implementation cycle for program activities while, simultaneously, Boeing supplied Langley and NASA Headquarters with very extensive supporting documentation, which detailed among other things the cost back-up data from the major subcontractors.

Scherer ascribed Boeing's excellent responsiveness during contract negotiations to the fact that NASA had predetermined the incentive features of the contract in the Request for Proposals. Moreover, the absence of a letter contract made it mandatory that negotiations be completed before actual work began, creating a sense of urgency for completing them as quickly as possible.[25] Boeing's willingness to listen to and analyze NASA's requests paid off on May 7, 1964, when James E. Webb signed the document approving the Lunar Orbiter contract and making the program an official NASA commitment.

Lunar Orbiter was a second-generation spacecraft and the first new start in lunar exploration since the decision to attempt a manned lunar landing mission to the Moon. The program's objectives were straightforward: the implementation at the earliest possible date of simple, reliable engineering measurements to determine the soundness of the

---

[25] OSSA Review, Lunar Orbiter Program Status Report, March 26, 1964, pp. 1-2.

spacecraft's design and the acquisition of scientific data about the Moon and its environment.[26] This information would prove vital for the mission design activities of the Apollo Program. In every respect, therefore, the Lunar Orbiter Program must be viewed as a direct support activity in implementing the decision to land men on the Moon and return them safely to Earth.

---

[26] Plans for Lunar Orbiter Data Acquisition and Analysis, Lunar Orbiter Program Office, March 20, 1964, pp. 1-2.

CHAPTER V

IMPLEMENTING THE PROGRAM

Early Funding Considerations

The beginning of the Lunar Orbiter Program's next stage was hardly noticed in the turbulent atmosphere in which the U.S. space program existed at home and abroad. Congress was questioning NASA and JPL about apparent poor management in the Ranger Program, while the first manned Gemini flight, scheduled for launch late in 1964, was experiencing setbacks. Everywhere, it seemed, the critics of America's space exploration efforts were finding fault with NASA. They pointed to Soviet manned and unmanned space accomplishments and asked why the United States was not keeping pace. In the midst of these inauspicious circumstances, the fledgling Lunar Orbiter Program at Langley nevertheless got off to a promising start.

Four aspects of the new program became important during the twelve months that followed the signing of the contract: 1) funding; 2) spacecraft design; fabrication, testing, and integration with the launch vehicle; 3) mission design; and 4) the establishment of schedules and working relationships between the various NASA centers and the contractors. Once the definitive contract with Boeing had been approved, funding problems became more complex. They constituted one of the dominant constraints defining the flow of activities during the entire course of the program. A brief description of funding through the end of 1964 will illustrate the problem.

Beginning in February 1964 the Office of Space Science and Applications had decided to commit to Lunar Orbiter the full $20 million which Congress had appropriated for FY 1964 specifically for an orbiter. However, the negotiated contract of April 16 obligated NASA to provide Boeing with funds as it required them, if the contractor was to be held to the incentive provisions in the contract. This meant that NASA had to establish and maintain a minimum funding rate to avoid schedule lags. Although NASA committed the FY 1964 funds, the Lunar Orbiter Program faced a new situation in FY 1965, beginning July 1, 1964. During the contract talks Boeing had predicted an expenditure rate of $26.1 million for that fiscal year, but by May this sum had increased to $37.1 million.[1]

A detailed PERT revealed one reason for this sudden rise. It found that by compressing the development phase of the program, NASA could gain more time for the testing phase. Acceleration of development, however, would require a higher funding rate than Langley or Headquarters had originally anticipated.

---

[1] NASA, Office of Space Science and Applications, Memorandum, Subject: Lunar Orbiter Funding, POP-64-3, August 24,

Realizing this the Office of Space Science and Applications released a guideline of $31.5 million for FY 1965 to the Langley Research Center in the spring of 1964. Of this Boeing would spend $28.9 million. Langley, on the other hand, had requested $39.1 million, of which Boeing was to spend $37.1 million. OSSA preferred to remain conservative, waiting until Boeing could supply more accurate, concrete information on funding needs before making a decision to increase the funding rate. Oran W. Nicks, Director of Lunar and Planetary Programs within OSSA, felt that the Lunar Orbiter funding requirements could increase at an uncomfortably fast pace and thus compromise other projects within OSSA.

Costs data for the Lunar Orbiter Program during the first quarter of the project, ending June 30, 1964, revealed that actual costs had exceeded estimated costs by $1.1 million. The estimated costs had been made by the Boeing Company on April 30, and the difference between the two constituted an underestimate by Boeing of 45% for the quarter.[2]

Throughout the summer of 1964 the rate of expenditure at Boeing remained Langley's single greatest headache. This was almost entirely due to Boeing's failure to sign

---

[2] Lunar Orbiter Project Office, Langley Research Center, **Project Lunar Orbiter, Narrative Analysis**, August 14, 1964.

the two major subcontractors, Eastman Kodak and RCA, to definitive contracts. Floyd L. Thompson kept Nicks informed of the funding problem during the summer months, and in August Nicks requested Thompson to review the entire funding situation and its potential impact on other programs.[3]

The scope of the funding problem revealed the need for closer cooperation between Langley and NASA Headquarters. Both organizations sent representatives to an August 19 meeting at Langley to examine and resolve their differences and strengthen the coordination of policies pertaining to Lunar Orbiter.[4] At the meeting officials from the various Langley offices connected with Lunar Orbiter gave detailed presentations of their work and requested further support of clarification of policies pertaining to the program.

Headquarters people made it clear that they wished to establish much firmer ties with Langley to ensure a better request-response relationship throughout the program. Langley people expressed concern that they had had to make decisions without the help of such useful tools as complete monthly funding reports from Headquarters which they could

---

[3] Memorandum from Oran W. Nicks, OSSA, to Floyd L. Thompson, Director of the Langley Research Center, August 20, 1964.

[4] Minutes of Lunar Orbiter Program Funding Meeting, Langley Research Center, August 19, 1964.

use to gauge their expenditure flow.[5]

Another pressing matter aired at the meeting was Langley's desire to fund Boeing three months in advance. This would allow enough flexibility to keep hardware procurement from falling behind schedule. But, because of the acceleration of development during the tight money situation in FY 1965, Langley's request appeared to be out of the question. Even with the present funding plan, funding to Boeing tended toward a minimum below which it could not go without precipitating serious schedule changes.

Langley and Headquarters officials decided to establish a minimum level for total expenditures at $41 million for fiscal 1965.[6] Cost reduction appeared unlikely in every program area except the Air Force Support Services at the Boeing Company. Here, according to Nicks, the very high projected cost figure of $2.45 million for FY 1965, which Langley's August Program Operating Plan had forecast, might be subject to reduction. In FY 1964 the U.S. Air Force had charged NASA an expensive 6% of Langley's combined contract costs as the fee for its support. NASA wanted the more reasonable rate of 1% to 2% which it received from the Navy and the Army for their various support services.

---

[5] _Ibid._
[6] _Ibid._

Nicks maintained that if NASA could obtain a figure of 1.5% of the Lunar Orbiter contract costs for FY 1965 as the rate of charge for USAF support, then it could alleviate some of the financial pressure which limited the flexibility of Lunar Orbiter funding in the coming fiscal year.[7] This new arrangement would have to be worked out with Air Force representatives.

Meanwhile the participants in the August 19 funding meeting agreed that no contract changes would be made if the changes would increase funding above the FY 1965 guidelines or above those laid down in the Project Approval Document or above the total program guidelines, unless the Lunar Orbiter Program Office in Washington had subjected the proposed changes to the most thorough scrutiny.[8]

The fact that the bulk of the procurement and development expenditures would come in FY 1965 further clouded the Lunar Orbiter funding situation. This reality placed a strict constraint on administration of the incentive contract with Boeing; it also prompted Langley Director Floyd L. Thompson to comment that, "if we aren't prepared to play table stakes, we shouldn't be in the incentive poker

---

[7] Memorandum from Oran W. Nicks, Director of Lunar and Planetary Programs, to the Director of Program Review and Resources Management, August 21, 1964.

[8] Minutes of Lunar Orbiter Program Funding Meeting, August 19, 1964

game."[9] To this Scherer added that, "when the government asks a contractor to assume the risk of an incentive contract, it must assume itself the responsibility for funding the contractor as he needs it."[10] He named the figure of $41.8 million as the rock-bottom minimum for the program in FY 1965 and stressed that any slip below this would cause schedules to lag and force basic alterations in the contract.

Lunar Orbiter funding became very tight in September at the time when Boeing was beginning to negotiate final contracts with Eastman Kodak and RCA. Langley informed NASA Headquarters that Boeing had received quotations from Eastman Kodak and RCA and, starting on September 14, would begin contract negotiations.[11] The original costs for the photographic system, which Boeing had quoted to Langley officials, proved to be much lower than the price at which Eastman Kodak was willing to deliver the subsystem for the spacecraft. This, in turn, had slowed contract talks between the two firms.

Scherer's main concern about the funding situation centered upon his recognition that to allow the program

---

[9] Memorandum from Lee R. Scherer to Oran W. Nicks concerning Lunar Orbiter FY 1966 Funding, September 4, 1964, p. 2.

[10] Ibid.

[11] Project Lunar Orbiter, Narrative Analysis, September 4, 1964.

to fall behind schedule because of too stringent funding would be tantamount to erasing the advantages of the incentive contract. If NASA induced the contractor to lose confidence in the contract because of a necessity to renegotiate part or all of it because of NASA niggardliness, then the program's overall success would be jeopardized. But NASA Headquarters remained steadfast in its retention of the $41.8-million FY 1965 funding minimum, even though Langley had called for $45.9 million.[12]

The growing seriousness of this problem brought Headquarters and Langley officials together on September 9. They established a new funding level based upon the increased requirements of Lunar Orbiter. This raised the original $94.6 million figure for the FY 1965-FY 1966 period to $105 million.[13] The new ceiling offered Langley greater flexibility and reassured the Lunar Orbiter Program Office in Washington that the incentive provisions of the Boeing contract would be maintained.

Both Langley and Headquarters concurred in the policy of holding all contract and schedule changes to the barest minimum. Moreover, both undertook studies of their opera-

---

[12] Memorandum from Scherer to Nicks, September 4, 1964.

[13] Memorandum from Homer E. Newell to Floyd L. Thompson, Subject: Guidelines for Lunar Orbiter Project, October 22, 1964.

tions to determine where costs might be reduced, and by the end of 1964 they had succeeded in pinpointing several ways to save more money. Scherer summarized the areas where cost reductions seemed most feasible and sent a report to Clifford H. Nelson at Langley at the end of December.

Boeing Negotiations with Subcontractors

Boeing satisfactorily completed technical negotiations with the Eastman Kodak Company by September 14, but cost negotiations became protracted. Eastman Kodak submitted a proposal of $27.1 million to Boeing, and this was substantially higher than the Boeing estimate of $19.3 million.[14] By October 6 the Langley Project Office realized that cost overruns for the spacecraft would be in the areas of procurement and the major subcontracts. Boeing resumed negotiations with Eastman and completed them by October 28. The Eastman contract would cost $22.4 million, which was still higher than the original Boeing estimate.[15] This meant that Boeing had already overrun the original contract by approximately $11.91 million: $3.07 million for procurement, $3.3-million difference

---

[14] *Project Lunar Orbiter, Narrative Analysis*, September 14, 1964.

[15] Ibid., October 28, 1964.

105

between budgeted and negotiated costs of the Eastman Kodak contract, and an estimated $5.64 million between budgeted and proposed costs for the RCA contract.[16]

Although negotiations with RCA originally were to run simultaneously with Eastman Kodak contract talks, they were delayed until Boeing had finished with Eastman. Scheduled for late November, the RCA talks were pushed back to December, when Boeing and RCA finally began cost negotiations. By December 9 RCA had offered Boeing a proposal for the communications subsystem with a total cost of $20.795 million for the spacecraft equipment and $5.329 million for the ground equipment. The cost was $8.4 million over the original Boeing estimate of $17.726 million.[17] Boeing did not complete cost negotiations with RCA until January 15, 1965, and the final cost figure was $22.6 million, substantially higher than the $17.7 million Boeing estimate.[18] These subcontracts brought the total cost of the Boeing contract to approximately $94.8 million by February 8, 1965. Of this, $4.0 million was for authorized changes and $10.3 million for estimated overruns.[19]

---

[17] Ibid., December 9, 1964.
[18] Ibid., January 25, 1965.
[19] Ibid., February 8, 1965.

NASA Cost-Reduction Efforts

Faced with the necessity to increase the rate of nding during the development and testing phases of the Lunar Orbiter Program, both the Langley Lunar Orbiter Project Office and the Headquarters Program Office initiated policies to reduce unnecessary costs wherever possible.

Learning from the Boeing-subcontractor negotiating experiences, NASA Headquarters and Langley continued to pursue the policy of keeping contract changes to an absolute minimum. The funding experiences of the second half of 1964 had made the managers of the Lunar Orbiter Program very cost conscious. The frequent meetings to discuss funding problems had improved communications between Langley and NASA Headquarters while they had also fostered a keen awareness by Boeing and NASA management of the implications and pitfalls in the Lunar Orbiter contract.

Besides the strictest limitations on changes, Lunar Orbiter could be spared undue expenses in another specific area: the planned need for redundant spacecraft to back up each flight spacecraft in the event of a failure before the launch. Originally the plans had called for the backup spacecraft, but after extensive consideration the Project Office at Langley concluded that direct substitution of one spacecraft for another between two launch windows, should the first spacecraft fail, was highly unlikely since the failure would probably necessitate an investigation of the other spacecraft.[20]

In addition to this, storage problems at Cape Kennedy and the necessity of maintaining the back-up spacecraft in mission-ready condition during preparation of the flight spacecraft presented no real guarantee of mission success but added extra costs to the program. Indeed the whole philosophy of spacecraft substitution seemed questionable, especially in a situation where every dollar counted. Scherer pointed out to Nelson in a memorandum that the earlier Pioneer and Surveyor programs had originally made provisions for back-up spacecraft but had later eliminated them. The Lunar Orbiter Program, by doing the same, could save a substantial sum of money.[21]

Elimination of the need for back-up spacecraft was not the only way savings could be made. The spacecraft delivery schedule proved to be another item for cost reduction. The spacecraft were scheduled to arrive at the Cape Kennedy facilities more rapidly than they could be launched. They would require storage space there, and this was very limited. As planned, spacecraft #8, the

---

[20] Memorandum from Lee R. Scherer, Lunar Orbiter Program Manager, to Clifford H. Nelson, Lunar Orbiter Project Manager, Langley Research Center, December 31, 1964, pp. 2-3.

[21] Ibid., p. 3.

last flight spacecraft, would arrive a full six months before its launch date; this would require that a "baby-sitter" keep it company for that length of time, clogging vital test and storage facilities. Scherer maintained that if changes were made in the delivery dates of the fifth through the eighth spacecraft, the storage vans and test teams could be reduced and money diverted for use elsewhere.[22]

One other item which Scherer explained to Nelson was the possibility of reducing costs by economizing on redundant recording equipment which the Lunar Orbiter Program would employ at each site of the Deep Space Network to record incoming data from the spacecraft. Comparing data-acquisition requirements of the Mariner Program with those of Lunar Orbiter, Scherer pointed out that Mariner had only two recording apparatuses per site, one of which served as a back-up. The Lunar Orbiter Program planned to have three or more, which seemed to be wasteful redundancy. He suggested to Nelson that he review the program's needs for so much recording equipment and, wherever possible, reduce or eliminate unnecessary extra equipment.[23]

---

[22] _Ibid._
[23] _Ibid._, pp. 3-4.

If funding difficulties for FY 1965 placed a major constraint on initial program operations, they also enhanced the performance of each task force engaged in the program, and the process of overcoming them educated Langley and Headquarters management as well as Boeing officials about the increasing complexity of the whole undertaking. It was clear by the beginning of 1965 that Boeing had originally underestimated the costs of the major subcontractors. The delays in signing both Eastman Kodak and RCA had made themselves felt in the area of development and procurement. Indeed, throughout the program the photographic subsystem would remain the pacing item, arriving late and at the Cape Kennedy facilities rather than at Boeing. Fortunately for Lunar Orbiter, NASA and Boeing personnel successfully circumvented the problems caused by the tardiness in signing the subcontractors to final contracts.[24]

---

[24] Recorded interview with James S. Martin, former Lunar Orbiter Assistant Project Manager, Langley Research Center, July 7, 1970.

CHAPTER VI

THE LUNAR ORBITER SPACECRAFT

A General Description

Before surveying the design and development phases of the Lunar Orbiter Program, it will be useful to describe the spacecraft which Boeing built for Langley. In the final design the Boeing Orbiter weighed about 385 kilograms and was 1.7 meters tall and 1.5 meters in diameter at its base, without including the solar panels and the antennas. Structurally the spacecraft had three decks supported by trusses and an arch. On the largest deck the main equipment was mounted: batteries, transponder, flight programmer, photographic system, inertial reference unit (IRU), Canopus star tracker, command decoder, multiplex encoder, and the traveling-wave-tube amplifier (TWTA), together with smaller units. Four solar panels and two antennas extended from the perimeter of this equipment deck.[1]

Above it, the middle deck supported the velocity control engine (the 100-pound-thrust Marquardt rocket motor), the fuel tanks, the oxidizer tank for the velocity control engine, the coarse Sun sensor, and the micrometeoroid de-

---

[1] Space Division, Boeing Company, The Lunar Orbiter, prepared for Langley Research Center, revised April 1966, pp. 20-21.

tectors. Above this the third deck contained the heat shield to protect the spacecraft from the heat generated by the firing of the velocity control engine. In addition the four attitude control thrusters were mounted on its perimeter. This uppermost deck was part of the engine module, which could be detached for test purposes. Directly under the engine was the high-pressure nitrogen tank, which provided pressure to feed fuel to the velocity control engine and to operate the attitude control thrusters.[2] This tank was one of the critical units; if anything caused it to lose pressure, the spacecraft could not manuever, and an entire mission could be ruined.

These and other items of spacecraft equipment formed subsystems of the whole spacecraft system. Working together they performed the Lunar Orbiter mission. The Eastman Kodak photographic subsystem has previously been described.[3] Electrical power was provided by a power system which operated in two modes: 1) solar panels converted solar radiation into electric current, and 2) batteries powered the spacecraft systems for short periods of occultation from the Sun. In periods when the solar panels would receive radiation from the Sun, the power supply would

---

[2] Ibid.
[3] See Chapter III.

run from the panels through the output voltage regulator to the other spacecraft systems (mode 1). This happened for the major part of the mission. At the same time power generated by the panels would also be directed into the battery charge controller, and from there a charging current would flow into the batteries as they could accept it. When no sunlight fell on the panels, the batteries would supply power to the output voltage regulator, and this would direct its flow to the spacecraft subsystems (mode 2).[4] In addition the power system had regulators and controllers to reduce unusual fluctuations to a minimum and enough solar cells to allow micrometeoroid damage to some without dangerous reduction in the capacity of the solar panels to generate electricity.

The attitude control subsystem served as the navigator for Lunar Orbiter during an entire mission. Composed of Sun sensors, the Canopus sensor, the inertial reference unit, and the thrusters, the system controlled the spacecraft's attitude in space in reference to the Sun, the star Canopus, and the Moon. The Sun sensors would "see" the Sun, produce signals which activated the attitude control thrusters, and these would align the spacecraft's roll axis with the sun. Once this reference was established the spacecraft could manuever off the reference and the IRU would remember

---

[4] Boeing, *The Lunar Orbiter*, pp. 26-27.

the original reference. If the need arose to move the spacecraft back to that reference, the IRU would signal the thrusters to correct the attitude. However, the IRU simply remembered reference points; it did not establish them.

Attitude control was directed by the flight electronics control assembly (FECA) and the Flight Programmer, which received data from all sensors and then informed ground control monitors, who could update the Programmer for future attitude manuevers. The FECA and the Flight Programmer controlled the spacecraft's attitude around its X (roll), Y (yaw), and Z (pitch) axes by activating the thrusters. They also governed the orientation of the photographic subsystem's camera lenses in relation to the surface of the Moon. Commands from Earth would make the spacecraft rotate through an angle around each axis according to the task to be executed, and the outputs of the gyros in the IRU would tell the Flight Programmer when the new attitude had been achieved. The Flight Programmer would stabilize and maintain the spacecraft in the new attitude relative to the three reference directions, and the IRU would tell it when there was any deviation from the established attitude.[5]

---

[5] Ibid., p. 28.

The Atlas-Agena D launch vehicle placed all five of the Lunar Orbiter spacecraft in parking orbits around Earth. The Agena with the spacecraft would remain in the parking orbit until the time to begin the translunar trajectory manuever, in which the Agena would fire out of Earth orbit toward the Moon. Once the spacecraft separated from the Agena there remained the task of correcting its initial trajectory and then of deboosting it into lunar orbit. The velocity control subsystem held the responsibility for this task and had to execute any changes in trajectory and speed.

The heart of the system was a 100-pound-thrust rocket whose hypergolic fuel and oxidizer ignited when the Flight Programmer commanded the intake valves to open. A burn to change the spacecraft's velocity would then occur and continue until the valves closed. Duration of any burn would be determined by information from the accelerometers in the IRU compared with prestored data in the Flight Programmer. The rocket engine was gimbaled to provide thrust vector control in order to accomodate center-of-gravity offsets and thrust asymmetries. The IRU accelerometers provided inputs for thrust vector control, the purpose of which was to keep the thrust of the velocity control engine through the spacecraft's center of mass.[6]

A nominal mission would provide for two midcourse manuevers to bring the Orbiter's trajectory precisely in line with an imaginary point where it would be deboosted into orbit around the Moon. At this predetermined point the velocity control subsystem would fire to slow the spacecraft and allow it to go into an initial orbit around the Moon. Ground personnel would then check out the spacecraft's orbital behavior and its various subsystems before making any decision to transfer to another orbit. Once they found the spacecraft's subsystems to be operating correctly, they would make a decision to inject it into a photographic orbit.[7]

Receiving and transmitting data to and from the spacecraft was the job of the communications subsystem, many of whose components had been flight-proven in the Ranger and the Mariner programs. This complex assembly could operate in four modes: 1) tracking and ranging, 2) command, 3) low power, and 4) high power. The communications system could send and receive data simultaneously while also transponding velocity and ranging signals for the Deep

---

[6] Interview with Leon J. Kosofsky, former Lunar Orbiter Program Engineer, NASA Headquarters, Washington, D.C., July 1, 1970.

[7] Boeing, *The Lunar Orbiter*, p. 29.

Space Network's tracking system.

The spacecraft's low-gain antenna picked up all incoming signals from the NASA-JPL Deep Space Instrumentation Facility stations. Commands from DSIF were routed to the command decoder and stored. The spacecraft would transmit a command from Earth back to Earth for verification before ground controllers sent an "execute" command. Upon receiving the execute command the communications subsystem would advance stored commands from the decoder to the Flight Programmer to be carried out. Photographic data with performance, environmental, and telemetry data would be transmitted to Earth by the high-power mode.[8]

Photographic data were transmitted in a different way than telemetry data were. The spacecraft had two antennas that operated in the S-band at the frequency of 2295 megacycles. Normally, when photographic data were transmitted to the ground receiving stations, the communications subsystems operated in the high-power mode and transmitted via the one-meter-diameter parabolic high-gain antenna. Simultaneous transmission of photographic and telemetry data was carried out as follows:

---

[8] Ibid., pp. 30-31.

> The 50-bit/sec telemetry data train is phase modulated onto a 30-kc subcarrier, which is then combined with the video data that have been transformed to a vestigial sideband signal. That signal is created by amplitude modulating the data on a 310-kc subcarrier by means of a double balanced modulator. This suppresses the carrier and produces two equal sidebands. An appropriate filter is then superimposed on the double sideband spectrum, essentially eliminating the upper sideband.
>
> Since the missing subcarrier must be reinserted on the ground for the proper detection of the vestigial sideband signal, provision for deriving such a subcarrier signal is made by transmitting a pilot tone of 38.75 kc. That pilot tone is exactly one-eighth of the original 310-kc subcarrier frequency, and is derived from the same crystal oscillator. Multiplying the received pilot tone by 8 in the ground equipment provides a proper subcarrier for reinsertion.[9]

Lunar Orbiter photographic data were never encoded; instead, data were transmitted as frequency-modulated analog signals. All other data from the spacecraft were encoded and sent on the subcarrier frequency as described above.

The temperature control subsystem protected all of the spacecraft's other subsystems from the extreme temperature variations of the deep space environment. Heat from the Sun could warm external parts of the spacecraft to 120°C while areas not exposed to solar radiation would cool down to -160°C. These extremes were beyond the temperature

---

[9] Leon J. Kosofsky and G. Calvin Broome, "Lunar Orbiter: A Photographic Satellite," *Journal of the SMPTE*, Vol. 74, (September 1965), pp. 776-777.

levels which most components could endure. The temperature control system established an environment ranging from +2°C to +30°C for the operation of all subsystems. A few components were exposed to direct sunlight: the four solar panels, the two antennas, the bottom of the equipment deck. The solar panels were designed to withstand temperature variations of +120°C to -160°C without cracking or buckling from severe expansion and contraction over a long period of time.[10]

Beginning at the uppermost deck a heat shield insulated the spacecraft from the rocket engine's heat while the entire area down to the lower deck was enshrouded in a thin-skinned aluminized mylar and dacron thermal blanket that covered all equipment except the Canopus star tracker's lens, the camera thermal door, and the components mentioned above. The bottom of the equipment deck, which faced the Sun most of the time during all five missions, was coated with a special paint having a high heat emission-absorption ratio. Small electric heaters were installed on the spacecraft inside the thermal blanket to raise the temperature if it fell below +2°C. The arrangement maintained everything under the thermal blanket at an average temperature.[11]

---

[10] Boeing, *The Lunar Orbiter*, pp. 32-33.

[11] Kosofsky interview.

The photographic subsystem had the most rigid temperature restrictions. Film could withstand heat only up to about 50°C, and moisture in the photographic subsystem would condense below 2°C, fogging the camera's two lenses. Eastman Kodak designed the system to be biased cool and warmed with little electric heaters. The "bathtub" housing the system did not touch the equipment deck but was affixed by four legs. Heat transfer between the "bathtub" and the equipment mounting deck was largely radiative, making heat absorption and dissipation a slower, more even process.[12]

One other component of the temperature control system was added after the original design to protect the photo-subsystem. This was the camera thermal door. Thermal tests showed that, without any cover over the camera's lenses, the lenses would be more susceptible to extreme temperature variations and stray light leaks inside. The major purpose of the camera thermal door was to reduce or eliminate the possibility that through heating the lenses could expand and alter the focal length so that distortions would result in the photography. The door would also help to control the internal temperature of the photo-subsystem so that it would not become too cold during periods of occultation and allow moisture condensation on the lenses. The door was added as one of the last components of the space-

---

[12] _Ibid._

craft before final design configurations were fixed. It was not part of the Eastman Kodak camera subsystem, and Boeing took the responsibility of designing, fabricating, and testing it.[13]

Early Design, Fabrication, and Testing Problems

One of the first hardware items to cause Langley and Boeing concern was the velocity control engine. The Boeing Company had proposed using the same Marquardt 100-pound-thrust rocket motor that the Apollo Program was using in the attitude control system of the Command Module. Lunar Orbiter would use this rocket for velocity control. During preliminary testing for Apollo requirements, the Marquardt rocket developed problems which caused Lunar Orbiter Program officials to have second thoughts about it. On April 21, 1964, Captain Scherer, with members of his staff and representatives of the Project Office at Langley, visited Marquardt to determine the seriousness of the problems and their implications for Lunar Orbiter.

His group learned that the Apollo mission requirements called for the rocket to be used in a pulse mode. It would have to fire reliably in short pulses thousands of times during an Apollo mission in order to change the Command Module's attitude as desired. Testing showed

---

[13] Interview with Thomas R. Costello, July 9, 1970.

that the rocket was not firing correctly in the pulse mode. This, however, did not affect its use in Lunar Orbiter, because as the spacecraft's velocity control engine it would be fired only at specific times in a single-burn mode.[14] Despite this difference in use Scherer recommended that until the Marquardt rocket proved reliable for Apollo such alternatives as the JPL Surveyor vernier engine should be studied.[15]

The Marquardt rocket was not so critical to the program's mission as another piece of hardware: the photographic subsystem's velocity-over-height sensor (V/H sensor). It could not be replaced easily by another component of a different kind, and its function was critical to the performance of the photographic subsystem. An image tracker which scanned a portion of the image formed by the 610 mm lens, it compared outputs derived from successive circular scans to measure the rate and direction of image motion before taking a photograph.[16]

The limitations of the V/H sensor determined in part the parameters of any photographic mission. It had to determine precisely the image-motion compensation values

---

[14] Kosofsky interview.

[15] OSSA Review--Lunar Orbiter Program Status Report, May 5, 1964, p. 2.

[16] Kosofsky and Broome, "Lunar Orbiter...," p. 775.

for photography below 950-kilometer altitude, where the spacecraft's velocity relative to the Moon's surface would affect the ground resolution of all photography. Above 950 kilometers the image-motion compensation could be deleted without significantly affecting ground resolution. At that high or higher altitudes the ground resolution of the high-resolution pictures might be reduced from 20 to 3 meters, but the case would be altogether different in an elliptical orbit which brought Lunar Orbiter as low as 46 kilometers above the Moon's surface. At this low altitude the camera would have to compensate for image motion to avoid "smearing" in a photographic exposure.[17]

Kosofsky and Broome have detailed why the V/H sensor is vital to low-altitude photography:

> The performance required of the image motion compensation apparatus is particularly exacting in the case of the Lunar Orbiter's high-resolution camera, as can be seen from the following figures. The design exposure speed is 1/25 sec, because of the very low exposure index of the film used (Kodak SO-243 film, with exposure index about 3). The spacecraft's orbital velocity at the low point of the orbit is around 1.6 km/sec, so that it moves 64 m across the target area during an exposure. In order to achieve 1-m ground resolution, the uncompensated image motion must be no more than the scale equivalent of 0.6 m. The allowable error in image motion compensation is thus 1%, which must be allocated between the mechanical limitations of the

---

[17] OSSA Review -- Lunar Orbiter Program Status Report, July 7, 1964, pp. 1-2.

platen servomechanism and the errors in the information supplied to it by the velocity/height (V/H) sensor.[18]

Eastman Kodak held total responsibility for producing the photographic subsystem for Boeing. However, it subcontracted work for certain components of the subsystem to Bolsey Associates. One of these components was the V/H sensor. Although both Eastman Kodak and Bolsey had very qualified men to design and build the components, management of their operations did not always run smoothly and adhere to schedules, as will be discussed later.

Two other problem areas became evident by September 1964 when Boeing commenced tests on the thermal model of Lunar Orbiter. The first was an overload on the power system because of increased need for electricity during periods when the spacecraft could not use its solar panels. The Inertial Reference Unit placed the greatest demand on the power system, and tests revealed that a battery with a greater capacity was probably needed to meet the demand. Boeing and Langley engineers also examined the possibility of changing the orbit design to give the spacecraft a longer period of sunlight instead of having to go to a heavier battery.

Review of the power system difficulties and subsequent

---

[18] Kosofsky and Broome, "Lunar Orbiter...," p. 775.

findings showed that under the planned night flying conditions the Orbiter's 12-ampere-hour battery would require an excessive charging rate, approximately 4.5 amperes, to meet the power needs of the other spacecraft subsystems. This high rate could cause battery failure, and Boeing engineers had worked out three possible solutions: 1) install a heavier, higher capacity battery, 2) turn off some equipment during the night periods, and 3) increase the time of the spacecraft's exposure to the Sun by altering the orbital parameters to be approximately 1,850 kilometers at apolune and 46 kilometers at perilune. The third solution would affect the spacecraft's photographic capabilities because the increased period of orbit would necessitate a decrease in the spacecraft's orbital inclination to the Moon's equator.[19]

During the Lunar Orbiter Program's First Quarterly Review at the Langley Research Center Scherer pointed out that, "if the initial orbit [of Lunar Orbiter] is made elliptical with a higher apolune, the day to night ratio would be improved and could be used to solve the problem."[20] Langley and Boeing adopted the third solution after Thomas Yamauchi, head of Boeing LOPO's System Engineering Section,

---

[19] Office of Space Science and Applications, NASA, Summary of First Quarterly Review, August 26-27, 1964, p. 4.

[20] OSSA Review --Lunar Orbiter Program Status Report, September 1, 1964, p. 3.

had worked out the rationale for the orbit change. The change did not greatly affect photography and eliminated the need for a heavier battery.

The second problem concerned the spacecraft's fuel and oxidizer tanks, which Boeing was purchasing from the Bell Aero Systems Company. Off-the-shelf hardware developed for the Apollo Program, the tanks had failed to pass qualification tests because of repeated rupturing of their teflon bladders. These bladders held nitrogen gas under pressure, and it was apparently seeping through the thin-walled bladders and saturating the fuel for the velocity control engine.[21] The Lunar Orbiter Program required extra qualification tests of the tanks, but this threatened to triple their cost. Langley requested the Office of Advanced Research and Technology to review the problem of the tanks while it looked into possible alternative solutions.[22]

On August 26, 1964, the Langley Research Center held the First Quarterly Review of the program to discuss all known problems which had come to light since the Boeing contract had been signed. Boeing representatives summarized their operations for Langley and Headquarters officials on

---

[21] Costello interview.

[22] OSSA Review, September 1, 1964, p. 1.

the first day of the review and then devoted the second day to detailed presentations on specific areas of the program to NASA personnel working directly in each area.

The Lunar Orbiter Program Office rated Boeing's total performance as very good, but noted that Boeing had treated its relationship with the Eastman Kodak and RCA subcontractors superficially. No representatives from EK of RCA were present at the Langley review, and officials of the Lunar Orbiter Program felt that a Boeing-Eastman Kodak-RCA team presentation at subsequent reviews would be very desirable.[23] Boeing, of course, was still in the process of signing contracts with these two firms.

During the review NASA and Boeing people treated the technical problem areas very thoroughly and discussed other difficulties related to spacecraft design and engineering. Boeing showed three more areas where work was required to attain the maximum functional efficiency in the spacecraft's configuration. The first was the spacecraft weight, a factor limited by the lifting capability of the launch vehicle. Boeing was aiming for a 370-kilogram spacecraft after separation from the Agena and before any midcourse manuever. The preliminary Lunar Orbiter design had indicated a 390-kilogram spacecraft, but two major steps had

---

[23] Summary of First Quarterly Review, August 26-27, 1964, p. 1.

successfully reduced this figure. First, Boeing had decided
to use integrated logic circuits in the control assembly
electronics, since this would save some 6 kilograms over
the use of discrete parts and perform just as well. Second,
the need to use one-pound thrusters in the attitude
control subsystem to compensate for thrust vector misalignment was eliminated when Boeing engineers redesigned the
system.

Originally the attitude control thrusters had been
located on the solar panels to take advantage of the greatest
moment. However, a close reexamination of this design convinced Boeing and Langley engineers that controlling the
thrust vector through the spacecraft's center of mass would
be substantially more difficult with one-pound thrusters
located far out on the solar panels. Attitude changes
could be executed easily, but they would cause perturbations
in the spacecraft's thrust vector which would have to be
counteracted if the spacecraft were not to assume a slightly
altered trajectory each time the thrusters were fired. The
process of counteracting changes in attitude would require
considerable fuel consumption on a thirty-day mission.

Boeing solved this design problem by eliminating the
four thrusters on the solar panels together with all of the
plumbing necessary to get gas out to them. This reduced
weight and the quantity of attitude control gas. Next the

velocity control rocket was gimbaled. The change required addition of two gimbals, their actuators, and bearings, but now the rocket's nozzle could be moved to compensate for any perturbations caused by the attitude thrusters. This resulted in a weight saving of about 3 kilograms. The attitude control thrusters were half-pound thrusters located at the perimeter of the heat shield. They were coupled so that when one of the four fired in one direction, its opposite number would fire in the opposite direction with the same amount of thrust for the same duration, changing the spacecraft's attitude without affecting the thrust vector.[24] This design change brought Lunar Orbiter's overall weight at the time of the Langley review to approximately 382 kilograms.

The participants of the review also tackled the problem of the Marquardt rocket motor, specifically the weight of the rocket's propellant versus the transit time from the Earth to the Moon and the specific impulse required to make the injection into lunar orbit. If the spacecraft was to achieve an initial elliptical orbit of 925 by 46 kilometers, it would require a total velocity change of slightly less than 1,100 meters per second. This meant that an Orbiter

---

[24] Costello interview.

[25] Summary of First Quarterly Review, August 26-27, 1964, p. 3.

weighing about 370 kilograms at separation from the Agena would require a specific impulse of 290 seconds. The Marquardt rocket, which had yet to pass qualifying tests for the Apollo Program, might not be able to achieve this high a specific impulse. (Although specific impulse is expressed in seconds, it is not a measure of duration. It is a measure of efficiency and indicates the thrust a rocket can provide at a certain rate of fuel consumption per second.) One possible solution to the problem, if the specific impulse of the rocket proved indeed too low, was to reduce the total impulse and alter the spacecraft's trajectory in order to place it in a more convenient initial elliptical orbit before transfer to final orbit.[26]

After reviewing the Marquardt rocket, the participants of the First Quarterly Review took up the examination of the last major problem to be considered at that time: Could the photographic system withstand the intense vibrations of the launch? The Eastman Kodak Company claimed that the vibration test levels were too high and that flight data on the launch vehicle did not warrant the high levels which Boeing had stipulated in its Environmental Criteria document. Boeing and Langley Lunar Orbiter Project Office people decided to reexamine the flight data of the Atlas-

---

[26] Ibid., p. 4.

Agena launch vehicle before making a decision on Eastman Kodak's complaint.[27]

This action ended the intensive two-day review of the program's major problem areas, and work proceeded. Two months later another review convened, and still more technical and engineering problems surfaced. They did not, however, threaten the comprehensive progress of the program toward its goals.

---

[27] Ibid.

CHAPTER VII

BUILDING THE SPACECRAFT: PROBLEMS AND RESOLUTIONS

Experiments for Lunar Orbiter

The Lunar Orbiter spacecraft was designed not only to take photographs but also to carry out three non-photographic experiments. A summary of these experiments will help to explain the direction of program thinking on scientific investigations of the lunar environment and show how the experiments presented problems for the total spacecraft configuration. The requirements of the Apollo Program and the weight limitations of the Agena rocket restricted the scientific payload of Lunar Orbiter to four experiments: photography, selenodesy, micrometeoroid, and radiation.

During the period in which the Request for Proposals was being prepared, the Office of Space Science through its Space Sciences Steering Committee evaluated the kinds of experiments which would be most useful to the scientific investigation of the Moon as well as to immediate NASA objectives. The major work of this evaluation fell to the Planetology Subcommittee.[1]

---

[1] See Minutes of the Planetology Subcommittee of the Space Sciences Steering Committee in the NASA Historical Office Lunar Orbiter History files. The meetings of the Subcommittee were conducted periodically during the entire course of the Lunar Orbiter Program.

The Subcommittee narrowed the field of experiments to be included on Lunar Orbiter early in the program's history. It found that one indispensable experiment the program should conduct was the recording of selenodetic information by tracking the spacecraft. The spacecraft would carry a transponder which would provide range and range-rate data, a necessity for mission control. Analysis of the data would establish a profile of the spacecraft's orbital behavior over a thirty-day period and longer. At a meeting of the Planetology Subcommittee on September 24, 1963, Gordon MacDonald of the University of California at Los Angeles had explained to Lunar Orbiter Program officials why the data were scientifically valuable as well as indispensable for the safety of the spacecraft on the first and subsequent missions.

He stated that if the Orbiters were to be flown in a low elliptical orbit around the moon, it would be mandatory to track the spacecraft on the first mission and determine its behavior by accurate measurements.[2] A selenodesy experiment which could record data for a period of at least sixty days at an altitude of 256 kilometers above the Moon on the first mission could sufficiently confirm the safety of putting subsequent Orbiters into orbits which would go as low as 32 kilometers above the Moon. Moreover,

---

[2] MacDonald's words understate the significance of the selenodetic data which the five Lunar Orbiters eventually gave. The discoveries made of the Moon's gravitational field by tracking the five spacecraft, especially Orbiter V, revealed the existence of large mass concentrations under the ringed maria on the nearside of the Moon. This orbital data enabled NASA scientists to construct a gravimetric map of the Moon's nearside in 1968, and the discovery of "mascons" by scientists of the Jet Propulsion Laboratory confirmed the presence of gravitational anomalies for both the Lunar Orbiter Program and the Apollo Program. The orbital behavior data of the five Lunar Orbiters convinced Apollo Program management it should redesign the Apollo 8 mission and plan an orbital mission for Apollo 10 rather than a landing, so that more precise tracking data could be gained before actually landing men on the Moon.

For a precise summary of the "mascon" phenomenon see: "Mascons: Lunar Mass Concentrations," by P. M. Muller and W. L. Sjogren of the Jet Propulsion Laboratory in Science, Vol. 161, No. 3842 (August 16, 1968), pp. 680-684. Refer also to the annotated bibliography in this history.

the selenodetic data gained in sixty days would be invaluable for the first Apollo lunar mission.[3]

Since its inception on May 4, 1962, the Lunar Sciences Subcommittee's Working Group on Selenodesy had developed information on lunar gravity and mass.[4] Originally the Group had provided major technical guidance for the Surveyor Orbiter Project at JPL. It made a timely contribution to Lunar Orbiter mission planning as a result of this earlier experience. The Group's chief concern was the design of the trajectory and orbits which the Lunar Orbiter would fly. Its work confirmed the limited extent of knowledge about the selenodetic environment and the potential hazards inherent in certain kinds of orbit designs. In its work it could little imagine the discovery in 1967 through the analysis of tracking data from <u>Lunar Orbiter V</u> of mass concentrations under the great maria of the Moon. The Working Group on Selenodesy provided MacDonald with a firm basis of fact for his argument that selenodetic data gathered by monitoring the Lunar Orbiter spacecraft in orbit would be very valuable for future orbital Moon missions.[5]

---

[3] Lunar Orbiter Discussion with Dr. Gordon MacDonald, September 24, 1963, Memorandum to the Record, October 2, 1963.

[4] Minutes: Working Group on Selenodesy, NASA Headquarters, May 4, 1962.

[5] <u>Ibid</u>.

A group led by William H. Michael at the Langley Research Center designed the Lunar Orbiter selenodesy experiment, and its efforts were richly rewarded by the data acquired during the five Orbiter missions.[6] Indeed, the selenodetic information that the program obtained substantially aided in extending the exploration of the lunar gravitational environment. When taken with the data from the five successfully landed Surveyors, these data provided the Office of Manned Space Flight very reliable, indispensable information for the Apollo Program.

In addition to selenodesy the Planetology Subcommittee selected two other fields of scientific investigation for experiments on the first five Lunar Orbiters which made up Block I of the program.[7] These were radiation and micrometeoroid flux in near lunar environment. The two experiments which Langley developed for the Orbiter were designed to measure the performance of the spacecraft as well as to provide useful data on potential hazards to manned missions to the Moon.

---

[6] Telephone interview with Dr. Samuel Katzoff, Langley Research Center, August 24, 1967.

[7] Originally the Lunar Orbiter Program had envisioned two blocks of spacecraft, but the lack of funds ended the development of more sophisticated Orbiters of Block II. A sixth flight spacecraft existed and could have flown after Lunar Orbiter V, but funds did not permit the flights.

The radiation experiment was designed by Dr. Trutz Foelsche and had two objectives as outlined by him:

> The principal purpose of the lunar orbiter radiation-measuring systems was to monitor, in real time, the high radiation doses that would accumulate on the unprocessed film in case of major solar cosmic ray events. In this way it would be possible for the mission control to minimize the darkening of the film by operational maneuvers, such as stopping the photographic operation and acceleration of development of the film in the loopers, and in case of more penetrating events, shielding the film in the cassette by the spacecraft itself and by the moon. Furthermore, the independent measurement of radiation doses would contribute to the diagnosis of film failure due to other reasons.
>
> A second purpose was to acquire a maximum amount of information on radiation on the way to the moon and near the moon, insofar as this could be achieved within the weight limitation of 2 pounds.[8]

The danger that the film could be damaged by solar radiation had Dr. Foelsche and Dr. Samuel Katzoff worried because the Eastman Kodak photographic subsystem provided only aluminum shielding at two grams per square centimeter at the film cassette and at two tenths of a gram per square centimeter in the rest of the system. Foelsche desired thicker shielding, but the contractors maintained that the film would be safe. The amount of shielding was a calculated risk, trading shielding weight against the probabilities of solar flare intensities.

---

[8] Trutz Foelsche, "Radiation Measurements in LO I-V (Period August 10, 1966 - January 30, 1968)," NASA Langley Research Center, paper to be presented at Manned Spacecraft Center Seminar, Houston, Texas, June 21, 1968, p.1.

Although he would have preferred to mount a more sophisticated experiment, Foelsche designed a measuring system to carry out the objectives described above, remaining within a one-kilogram weight limit. The system's sensors, their arrangement and shielding, the measuring principle and dynamic ranges were all developed at Langley. The Lunar Orbiter Project Office at Langley and the Boeing Company then determined the specifications for the hardware, and Texas Instruments built and calibrated the experiment.[9]

The micrometeoroid experiment was the last non-photographic experiment which the Planetology Subcommittee approved for the Block I Orbiters. Designed by Charles A. Gurtler and William H. Kinnard of Langley, it consisted of twenty detectors mounted around the middle deck of the spacecraft, outside the thermal blanket. Each detector consisted of a pressurized semicylinder with a pressure-sensitive microswitch inside. The cylindrical surface of the detector was 0.025 mm beryllium copper test material. Inside the semicylinder, gas pressure held the switch closed. When a puncture of the surface material occurred, gas would escape, opening the microswitch, which would register the puncture electrically. Whenever the condition of the

---

[9] *Ibid.* See schematic diagram on following page.

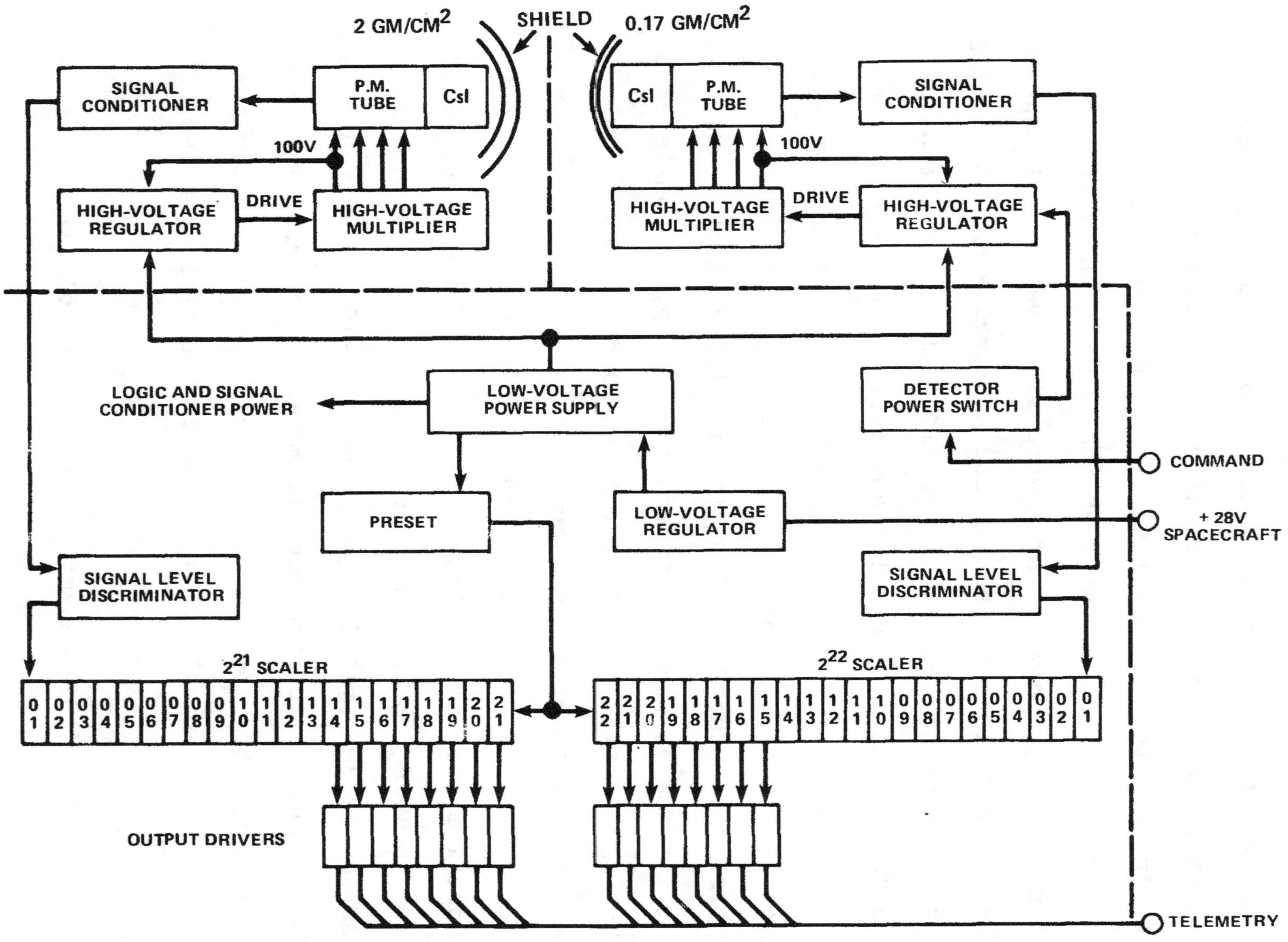

SCHEMATIC OF LUNAR ORBITER DOSIMETER SYSTEM

detectors was telemetered to Earth, any new punctures would be indicated and previously indicated ones would be verified (see diagrams on following pages).[10]

Gurtler and Kinnard presented their experiment to the OSSA Space Science Committee on October 5, 1964. After reviewing it, the Committee pointed out that the instrumentation was omnidirectional and limited in the quantity of data it could acquire. The Committee requested Gurtler and Kinnard to examine the kinds of similar instrumentation which the Surveyor and the Mariner C spacecraft had and to ask W. Merle Alexander at the Goddard Space Flight Center in Greenbelt, Maryland, for specific assistance in the further study of the experiment's requirements, since Alexander was the principal investigator for micrometeoroid instrumentation on these two spacecraft.[11]

In the end, however, Gurtler and Kinnard's experiment was implemented in the form originally presented to the Committee. While the instrumentation could provide only limited data, it had the advantages of simplicity and freedom

---

[10] C. A. Gurtler and Gary W. Grew, "Meteoroid Hazard Near Moon," Science, Vol. 161 (August 2, 1968), p.462.

[11] Memorandum from Dr. Homer E. Newell, Associate Administrator for Space Sciences, to Dr. Floyd L. Thompson, Langley Research Center, October 23, 1964.

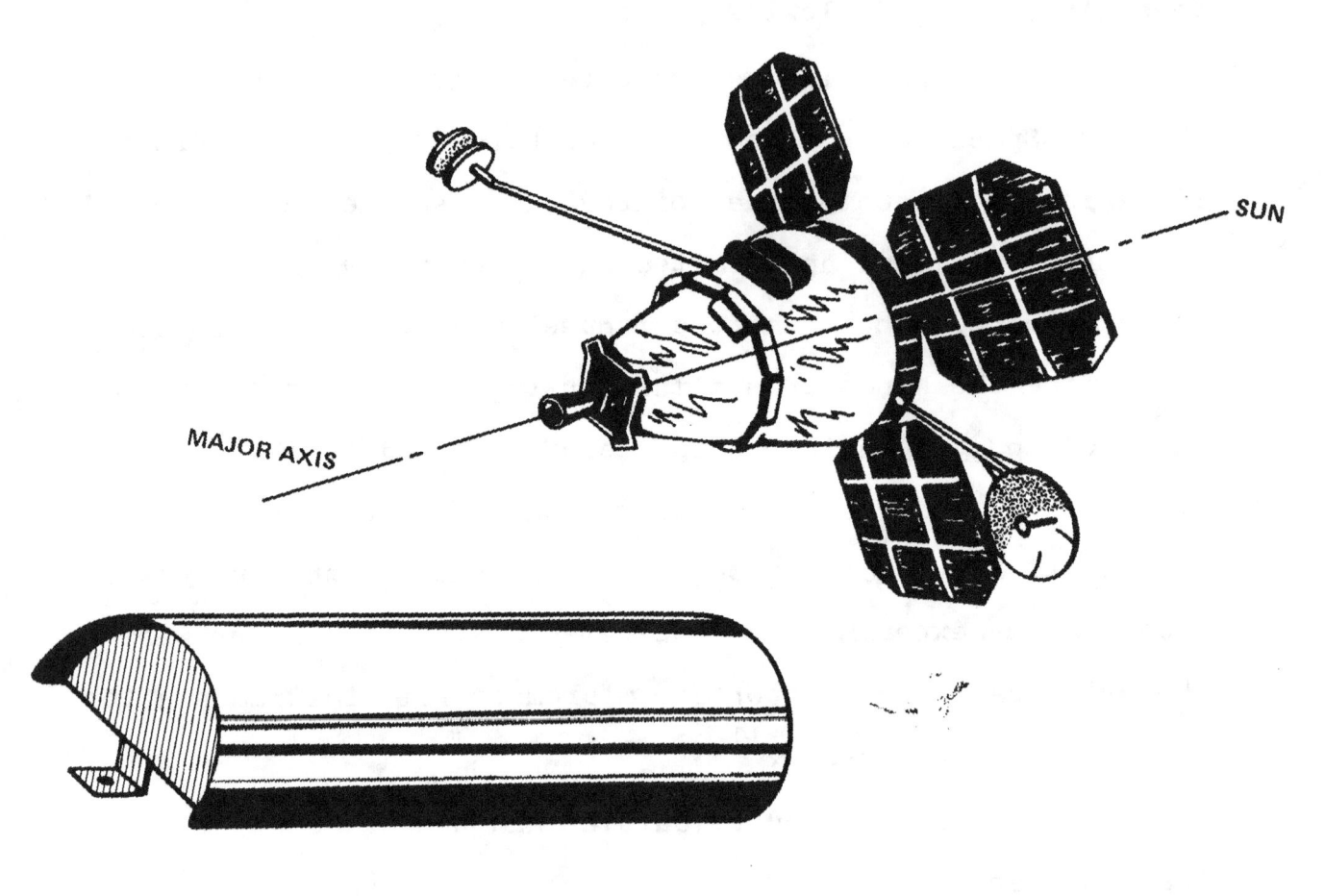

GEOMETRY OF METEOROID DETECTORS ON SPACECRAFT

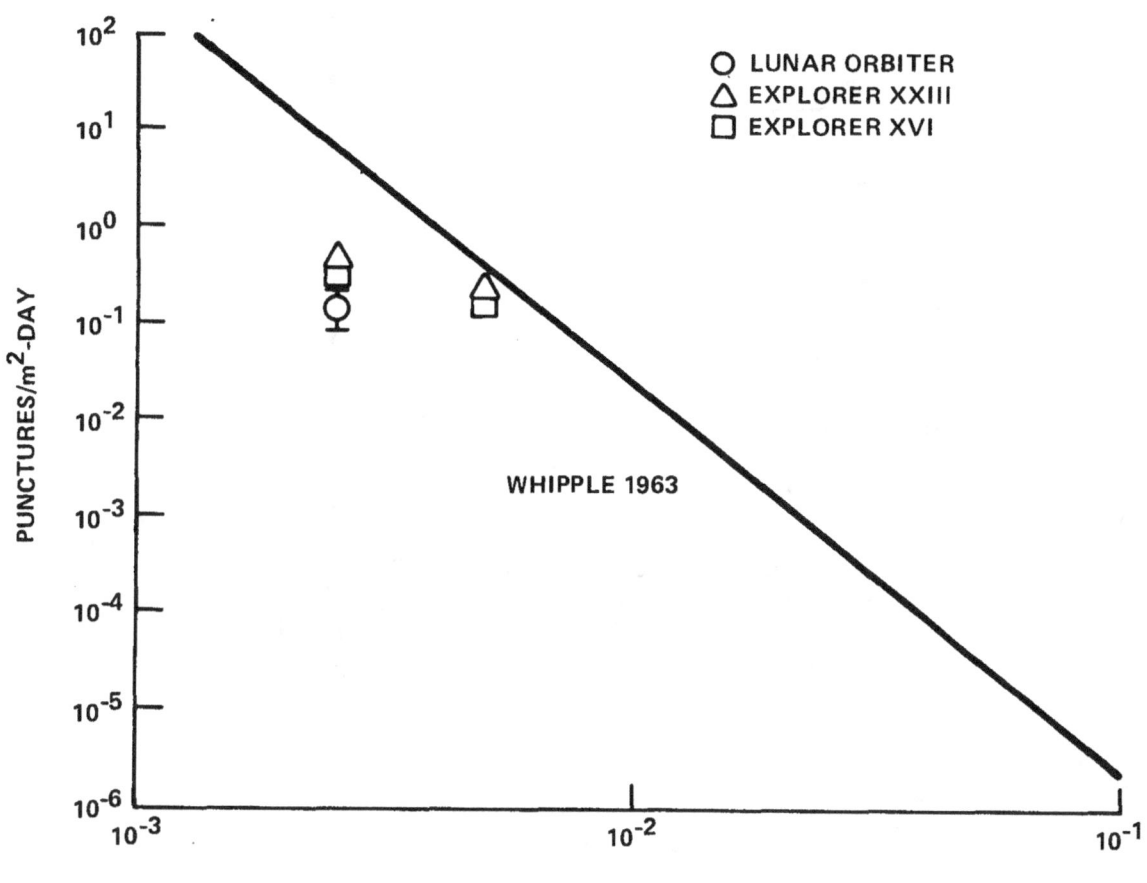

Puncture Rates for Beryllium Copper Pressurized-cell Detectors on Lunar Orbiter and Explorer XVI and XXIII and Comparison with Whipple's 1963 Prediction, Converted to Beryllium Copper.

MICROMETEOROID PUNCTURE RATES

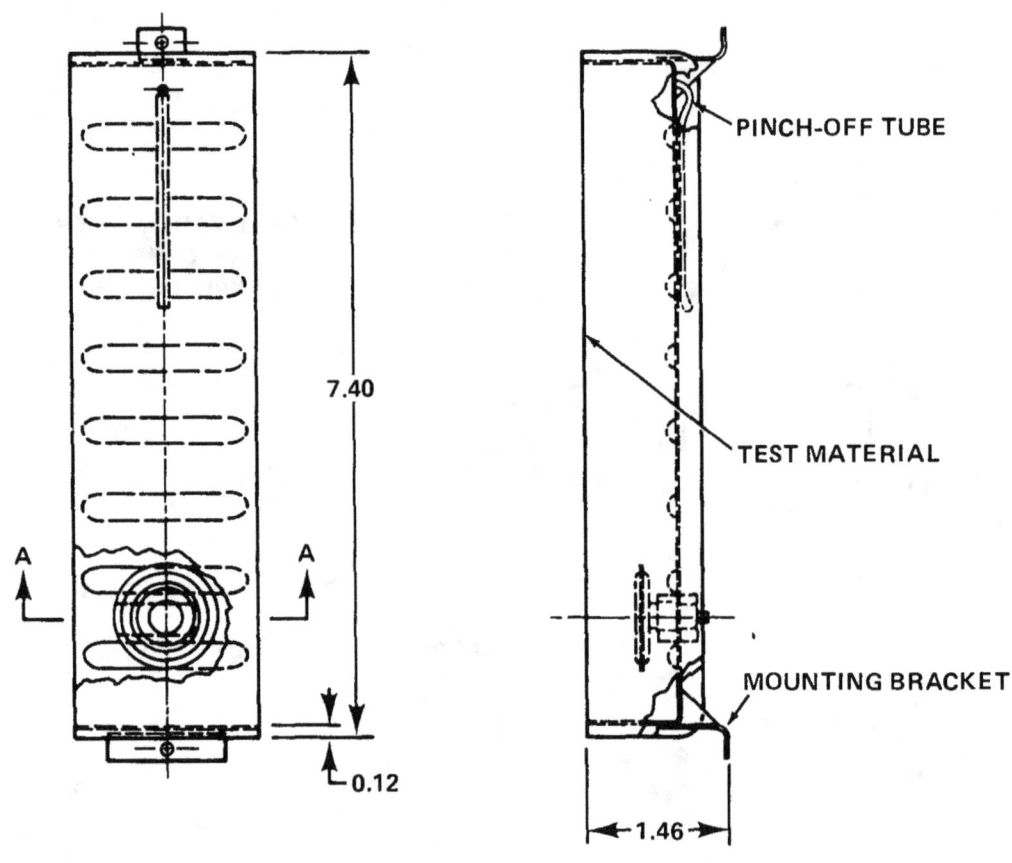
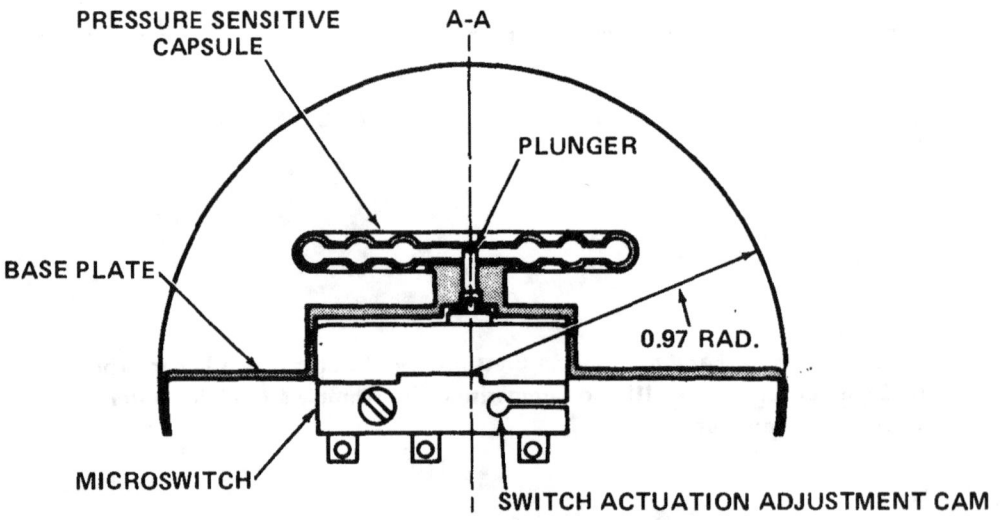

PRESSURIZED-CELL DETECTOR
(DIMENSIONS IN INCHES)

from ambiguity.

The photographic experiment, which constituted the major means of implementing the program's objectives, has been discussed previously and will be referred to during the course of this narrative as the need arises.

Other Potential Experiments

Although the Block I spacecraft carried only the four experiments described above, the Lunar Orbiter Program Office was planning a greater number of more sophisticated scientific experiments for the Block II Orbiter. They included: 1) a gamma ray experiment to determine the presence and relative abundance of natural, long-lived radioisotopes on the surface of the Moon; 2) an infrared experiment for mapping the lateral variations in the Moon's surface temperature; 3) a bi-static radar experiment for determining the average radar cross-section, surface roughness correlation functions, altitude measurements, reflectivity, and the dielectric properties of the lunar surface; 4) a photometry/colorimetry experiment to determine variations in the photometric function and the color of lunar surface materials; 5) a radiometer experiment for measurement and determination of lunar surface thermal gradients; 6) an X-ray fluorescent experiment to detect the relative abundance of iron and nickle on the Moon's surface; 7) a solar plasma experiment to study the spatial and temporal flux variation and energy distribution of low-energy protons and electrons of the plasma; 8) an experiment to investigate the magnetic field in the vicinity of the Moon; and, finally, 9) a lunar ionosphere experiment to determine the presence of a low-density ionosphere in the immediate vicinity of the Moon's surface.[12]

These experiments, spanning a wide range of scientific fields of investigation, demonstrated that the Lunar Orbiter Program envisioned in a second block of spacecraft a series which would conduct primarily scientific investigations and not necessarily more photography of the lunar surface. NASA had already designated the Block I Orbiters for missions which would gather photographic data of the lunar surface vital for mission planning of the Apollo Program.

Moreover, the first Lunar Orbiters would explore some aspects of the Moon's environment and complement the work which the Surveyor spacecraft would carry out when they landed on the Moon. The Orbiter concept, expanded in a second series of spacecraft, could achieve major advances in knowledge about Earth's natural satellite, a philosophy consistent with the mainstream of thought in the Office of Space Science and Applications. However, lack of funds eventually precluded the Block II Orbiters and curtailed a

---

[12] Martin J. Swetnick, "Unmanned Lunar Scientific Missions, a Summary," November 17, 1964. Dr. Swetnick was a Lunar Program Scientist.

major U.S. scientific thrust in exploring the Moon.

Preliminary Mission Planning Activities

A third area of the Lunar Orbiter Program was mission design, and success in planning the missions to be flown depended heavily upon coordination among the various NASA and industry participants. Implementation of the planning activities depended upon the establishment of schedules for the program's various task groups; in turn these had to be integrated with one another to effect the timeliest utilization of information within each specific area of the Lunar Orbiter Program.

Although detailed consideration had been given to ways and means of utilizing NASA's capabilities to facilitate Boeing's work during the period of contract negotiation, the first major meeting to discuss actual schedules and working relationships convened on April 15, 1964, at the Langley Research Center. The meeting's purpose was twofold. First the participants from Headquarters, Langley, Lewis, JPL, and Boeing had to work out a basic agreement about the delegation of responsibilities which had not yet been assigned through any earlier agreements. This included tentative declarations by each party of its capabilities and limitations and what tasks each believed it could best perform to contribute to the success of the program. Secondly, the representatives of the various centers and the prime contractor had to agree upon the implementation of the decisions in the first area of agreement.[13]

Thomas Yamauchi of the Boeing Company began the talks with a presentation of a condensed project schedule and noted the time intervals in which Boeing would require trajectory information from the Lewis Research Center and JPL concerning the launch vehicle and tracking and data-acquisition needs. He outlined the kind of information which Boeing would require from each.[14]

Dr. Karl A. Faymon of Lewis responded by specifying approximately the times before each launch when Lewis could deliver various preliminary and final data on launch vehicle checkout and performance. He also explained the times at which Boeing would have to supply data to Lewis on launch constraints, detailed mission profiles, and updated weight estimates. The flow of information between Lewis and Boeing appeared not to present any serious problems at the time of the Langley meeting.[15]

While the job which Lewis would perform for Boeing

---

[13] Memorandum to the Record, Summary of Lunar Orbiter Trajectory Meeting, Langley Research Center, April 15, 1964 (document dated April 17, 1964).

[14] Information was not enumerated in the document.

[15] Summary of Lunar Orbiter Trajectory Meeting.

and the Lunar Orbiter Program concerned hardware, the role which the Jet Propulsion Laboratory and the Deep Space Network would perform was much more complex. The services which JPL and the DSN would render fell into two categories: flight programs and tracking and data acquisition. Both required different kinds of organization. JPL had already committed the Deep Space Network facilities which the Lunar Orbiter Program would require, and these and their operation came under the auspices of the NASA Office of Tracking and Data Acquisition (OTDA). There was little trouble here between Langley and JPL.

The work which JPL flight programs manpower could reasonably render the Lunar Orbiter Program was another matter. Before JPL could do anything, it had to know the amount and kind of resources which Langley desired that JPL commit to Lunar Orbiter. In this case JPL's ability to commit the resources depended upon its commitments to other flight programs: Ranger, Surveyor, and Mariner. These programs were all funded through the Office of Space Science and Applications, and any decision about an increased work load for JPL would have to take them into consideration.[16]

---

[16] Letter, Rechtin to Emme, November 18, 1969.

When Langley had requested additional support from JPL on April 2, the request was not for work to be done by the DSN. It fell instead within the realm of flight programs, and JPL manpower was already spread thinly. On April 2 Langley had requested of NASA Headquarters that JPL take on the responsibility "for the programming of all operational computer programs, including reviewing the physical and engineering problems they represent, their mathematical formulation, and the formal requests for programming." This was not all. Langley wanted JPL to "make a definitive study of Lunar Orbiter tracking data requirements, including the accuracy of realtime trajectory determination, considering tracking sites, data types, sampling rates, data noise biases, site errors, etc."[17]

The Lunar Orbiter Project Office at Langley also wanted JPL to "check the Space Flight Manuever Specifications Tables; i.e., the guidance philosophy for midcourse, deboost, and retro firing, including numerical firing tables which will be used in DSN operations."[18] Boeing, at the same time, was to conduct a similar study of tracking and data-

---

[17] Memorandum from Floyd L. Thompson, Director of the Langley Research Center, to Homer E. Newell, Subject: Request for additional support for Lunar Orbiter from Jet Propulsion Laboratory, April 2, 1964.

[18] Ibid., p.1.

acquisition requirements and was to review all JPL support work. When Floyd L. Thompson had presented these expanded requests to Marshall Johnson, the Tracking and Data Systems Manager at the DSN, and Victor Clarke, also of JPL, they had reacted favorably but had stipulated that the Systems Analysis Section and the Computer Applications and Data Systems Section at JPL would require more manpower to perform the Lunar Orbiter work.[19] However, Johnson and Clarke were part of the DSN, not the JPL flight programs operation, and they were not in a position to commit non-DSN resources.[20]

At the April 15 Langley meeting JPL representatives proposed a multi-staged program to educate Boeing and Langley personnel about the capabilities of the DSIF and SFOF so that they, in turn, could use their manpower to perform the flight operation tasks necessary to the preparation and execution of each mission. JPL also suggested that Boeing set up a computer facility to "resemble" the Space Flight Operations Facility and run its own programming while having a private contractor check it independently.[21]

---

[19] Letter, Rechtin to Emme, November 18, 1969.
[20] Ibid.
[21] Summary of Lunar Orbiter Trajectory Meeting, pp. 1-2.

Langley and JPL proceeded to work out a compromise agreement to facilitate the timeliest integration of schedules. The actual problems of mission design and orbit determination remained in the hands of the Lunar Orbiter Project Office, specifically under the direction of William J. Boyer, the LOPO Operations Manager, and John B. Graham, in charge of operations integration.

Robert J. Helberg at Boeing assigned Thomas Yamauchi to coordinate mission planning with the LOPO at Langley. On June 10, 1964, a major meeting convened at NASA Headquarters to review the status of Yamauchi's work, the proposed first mission, and the technical problems which placed constraints on the design of that mission. It had become apparent to Scherer, Kosofsky and Swetnick of the Headquarters Program Office that a dichotomy existed between the requirements of the short-term photographic mission and the extended selenodetic mission of the spacecraft. This dichotomy affected design of the attitude control system, since its performance could determine the orbital parameters of the spacecraft during the long-life mission which was to last about one year after termination of photography and readout.[22]

Scherer outlined the first tentative Lunar Orbiter

---

[22] Memorandum to the Record from Martin J. Swetnick, Subject: Summary Minutes, Lunar Orbiter Meeting at NASA Headquarters, June 10, 1964, document dated June 22, 1964.

mission to the participants of the meeting as an introduction to the areas of difficulty. Mission A, as it was later called, would inject an Orbiter into a nearly circular orbit approximately 925 kilometers above the Moon with an inclination of 21° to the lunar equator. The orbit was then to be changed to an ellipse ranging from 925 kilometers at apolune to 46 kilometers at perilune, because this would be most satisfactory for high- and medium-resolution photography.[23]

Dr. Gordon MacDonald of UCLA, a member of the OSSA Planetology Subcommittee, expressed some doubt about the safety of the spacecraft at such a low perilune over a period of one year. His reasoning was based upon the fact that the attitude control system, as it was then designed, would cause periodic perturbations in the orbit by repeated firing of its thrusters. (At this time the Orbiter had one-pound thrusters located at the tips of the solar panels. When fired they would change the spacecraft's attitude, but they would also cause some oscillations in the solar panels and would affect the spacecraft's thrust vector.) This could cause a three-meter change in the perilune per orbit, according to MacDonald. A Boeing study that Yamauchi had directed substantiated his conclusion. The change would be

---

[23] Ibid.

too great for the spacecraft's velocity control subsytem to handle over the long run and could jeopardize the extended mission. MacDonald suggested that Boeing make a detailed analysis of the attitude control subsystem and its effects on the velocity and thrust vector control.

The members of the meeting agreed that Boeing should examine the following questions:

1. What dead zone can the Lunar Orbiter attitude control system accept on an extended mission?

2. What will be the effects of the control jets on the motion of the Lunar Orbiter?

3. Can the impulses on each control jet be measured and counted, even during the time the spacecraft is not within line of sight telecommunications to earth?

4. What possible effects can an imbalance, such as the high gain antenna on the end of a boom, have on the attitude of the Lunar Orbiter over an extended lifetime mission?

5. Is it possible to modify the design of the attitude control system to operate coupled pitch and yaw jets?[24]

Following the meeting, the Boeing Company went to work on the design of the attitude control subsystem, and by the First Quarterly Review at the end of August, the spacecraft design was beginning a three-stage metamorphosis which would result in its final configuration in the spring of

---

[24] *Ibid.*, p.5.

1965.[25] The metamorphosis through April 1965 can be briefly summarized.

Initially the spacecraft had a photographic subsystem housed in a barrel-shaped "bathtub." The attitude control thrusters were located at the periphery of the solar panels with requisite plumbing to feed gas to them from storage tanks in the engine module. At stage two the spacecraft had a more efficiently shaped "bath tub" with a flat bottom for better thermal control. An arch from the equipment deck to the middle deck had been placed over the photographic subsystem to add strength, and the structure of the velocity control subsystem had been changed. However, the attitude control thrusters still remained at the tips of the solar panels.

In the third stage stage of the metamorphosis the velocity control engine had been gimbaled, the change reducing its fuel requirement and allowing more room for the nitrogen tank to fit down into the center of the engine module. The attitude control thrusters had been reduced from one-pound to one-half-pound thrusters, and they had been relocated on the periphery of the uppermost deck of the engine module. They had also been coupled, and the need for the plumbing to carry gas to the tips of

---

[25] Summary of First Quarterly Review, August 26-27, 1964.

the solar panels had been eliminated. The omni-antenna boom had been strengthened, and the micrometeoroid detectors had been placed around the middle deck.[26]

These changes raised technical design problems, but they also affected preliminary mission planning activities--as did the working arrangement established between Langley and JPL. At the beginning of July 1964 officials from the two centers worked out the details for educating selected Langley and Boeing personnel in mission analysis, programming standards, and the review of existing programs that might benefit Lunar Orbiter. Training began on July 15 and afforded the Lunar Orbiter Program the opportunity to solve its own problems of analysis without unduly taxing JPL manpower.[27] Boeing was very willing to learn from JPL, a fact which facilitated the implementation of the Langley-JPL working agreement and, indeed, overall mission success in the program.

Testing Procedures and Program Reviews

One important feature of the Lunar Orbiter spacecraft was that its design did not rely heavily upon

---

[26] OSSA Review--April 13, 1965, p. 1. See diagram on the next page.

[27] Memorandum from Lee R. Scherer, Lunar Orbiter Program Manager, to Oran W. Nicks and Edgar M. Cortright, Subject: Immediate need for JPL support for Orbiter, July 10, 1964.

LUNAR ORBITER SPACECRAFT

redundant subsystems or components. Moreover, although the subsystems were integrated, they were not heavily interdependent and could function more independently of each other than the subsystems could in such spacecraft as Mariner. This design concept reflected Boeing's long-standing traditions in aircraft, and it paid off handsomely.

The testing philosophy of the Lunar Orbiter was one reason the design proved to be so successful. Several kinds of tests and reviews were used in the program. First was the Preliminary Design Review, conducted by NASA and Boeing. This form of review was always held to check any specific technical area or major subsystem before a final decision was made to freeze the design. When agreement was reached, Langley gave Boeing permission to fix the design, and then both parties met to hold a Critical Design Review. In this review the item, whether a component or a major subsystem, was picked apart or passed as acceptable for fabrication and testing. If approved, the item was procured or fabricated, and after approval Langley tried to hold changes to an absolute minimum. During the fabrication stage, various forms of reviews took place until the item was completed and tested. At the completion point, a formal NASA

Acceptance Review was conducted.[28]

The Langley-Boeing testing procedure was aimed at making the first mission a complete operational success. The procedure played a vital part in the program and reflected the positive attitudes throughout the entire Lunar Orbiter Program team.

At the beginning of the whole testing sequence, all components of the spacecraft system went through a Flight Acceptance Test (FAT), which exposed them to "nominal"-- or expected --vibration, temperature, and vacuum conditions of operational environments. Three sets of each component were then divided into sets A, B, and C for more specific tests. Set A was used for qualification tests simulating overstress conditions. This kind of test was designed to push the component beyond expected endurance limits to determine what punishment it could actually withstand. Set B underwent reliability demonstration tests that simulated two real-time missions at the FAT level. Finally, Set C components made up subsystem assemblies that were tested and then

---

[28] Robert J. Helberg and Clifford H. Nelson, "The Lunar Orbiter -- An Integrated Design," paper presented at the XVIII International Astronautical Congress, Belgrade, Yugoslavia, September 27, 1967, pp. 607. Helberg was Assistant Division Manager—Spacecraft Systems, Space Division, The Boeing Company, and Nelson was Lunar Orbiter Project Manager at Langley Research Center.

integrated into a complete spacecraft (Spacecraft "C").

This first complete spacecraft system, minus the photographic subsystem, was subjected to compatibility tests with the Atlas-Agena launch vehicle; with the tracking and communications network at Goldstone, California; and with the Eastern Test Range tracking and communications facilities at Cape Kennedy.[29] The idea to test the spacecraft for compatibility with the DSIF facility at Goldstone had been suggested by JPL; Langley accepted it, and testing proved to be very useful in establishing biases between the Lunar Orbiter communications subsystem and the DSIF receiving station.[30] A test film was read out during dry-run exercises there to check the accuracy in the transmitting and receiving equipment.

Boeing built a total of eight Lunar Orbiter spacecraft for the program, including Spacecraft C. Following Spacecraft C came Spacecraft 1 and 2. Number 1 underwent qualification tests at spacecraft level while Number 2 was subjected to thermal vacuum tests for a period covering the duration of two missions. The other five Lunar Orbiters (3, 4, 5, 6, and 7) were put through Flight Acceptance Tests

---

[29] Ibid. See figure, Lunar Orbiter Test Program, on next page.

[30] Letter, Rechtin to Emme, November 18, 1969.

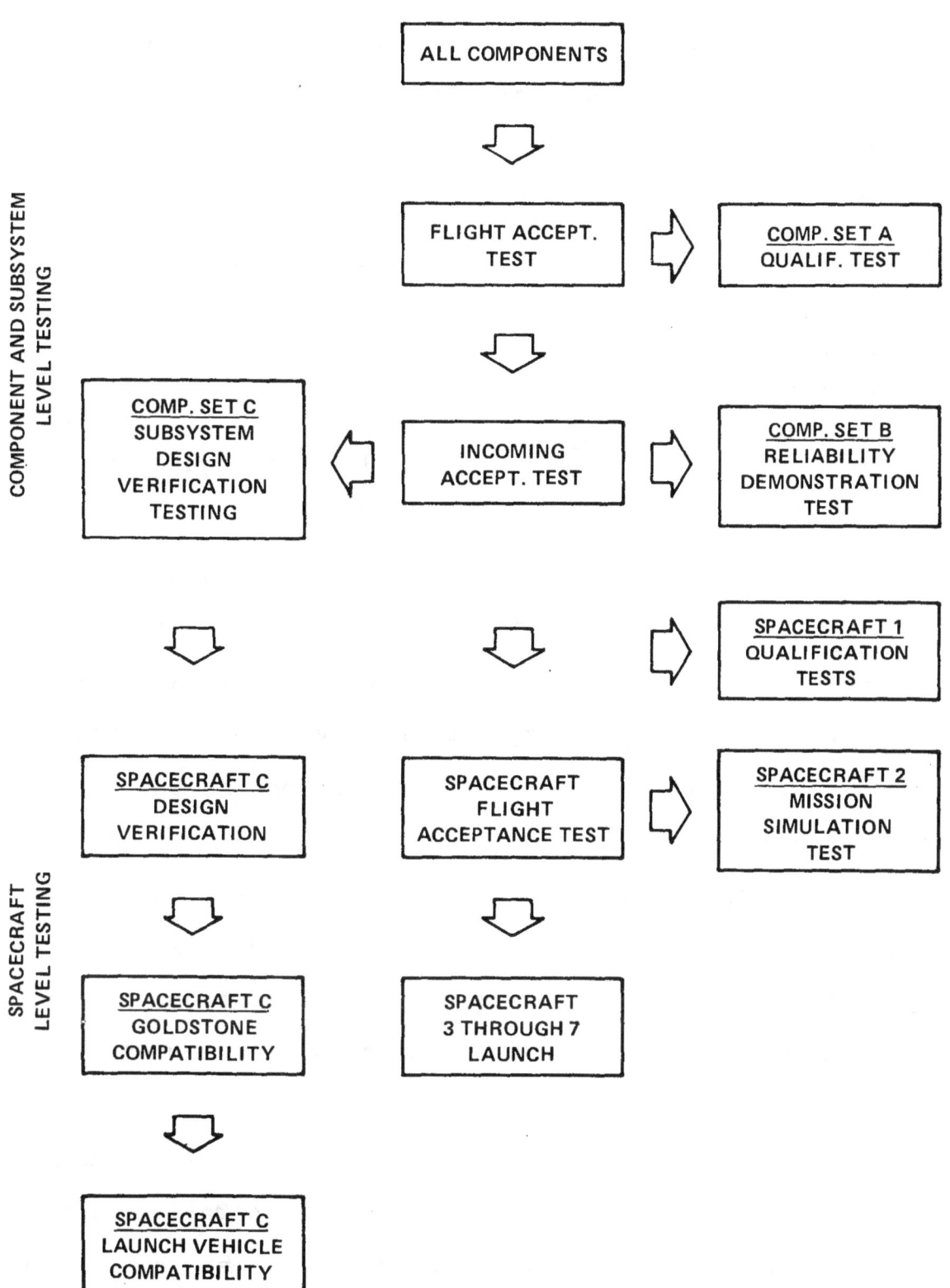

LUNAR ORBITER TEST PROGRAM

161

and then sent to the Eastern Test Range for their final checkout and launch. The chart below clarifies the sequence:

	Spacecraft Number							
	C	1	2	3	4	5	6	7
Lunar Orbiter		Ground test spacecraft		V	I	II	III	IV
Mission				E/5	A/1	B/2	C/3	D/4

Clifford H. Nelson pointed out to the participants of the XVIII International Astronautical Congress in Belgrade, Yugoslavia, that no serious problems or failures were experienced during all spacecraft-level tests in the program. This testified to the standards and the thoroughness which Boeing and Langley had used in testing at the component and subsystem level, and it also testified to the excellence of the spacecraft's design. Faulty equipment and poor designs had been effectively rooted out during the testing phase of the program when potential problems in subsystem integration had been exposed.[31]

More interesting, however, was the fact that Boeing and Langley had agreed early on testing in a parallel mode rather than in a series mode. Tight schedules and a spartan

---

[31] Helberg and Nelson, "The Lunar Orbiter -- An Integrated Design," p. 8.

economy were largely responsible for this. Thus, for example, the three sets of components (A, B, and C), Spacecraft 1 and 2, and the five Flight Spacecraft (3, 4, 5, 6, and 7) were tested in periods that substantially overlapped.[32] Ira W. Ramsey headed a team of men in the LOPO which was responsible for the entire Lunar Orbiter testing program and for the success of the parallel mode despite its inherent risk.[33]

Problem Areas: Last Quarter 1964 to First Half 1965

Several problem areas had developed by late 1964 which threatened the original schedules of the program. Some of these have already been mentioned. Two more are noteworthy, however. At the Lunar Orbiter Preliminary Design Review held at Boeing on October 27 and 28, 1964, the status of the micrometeoroid and radiation experiments had somewhat alarmed Israel Taback, the Langley Lunar Orbiter Spacecraft Manager, and Martin J. Swetnick, the Lunar Orbiter Program Scientist from NASA Headquarters. They learned that the instrumentation which Boeing proposed to procure for the two experiments by letting bids to Space

---

[32] Interview with Gerald Brewer, Chief of Mission Assurance, Lunar Orbiter Project Office, Langley Research Center, July 18, 1967.

[33] Refer to Project Organization Chart in Appendixes.

Technology Laboratories or Texas Instruments, Inc., did not meet the actual specifications in the experiments document. Indeed Taback and Swetnick felt that even the specifications document which Boeing had drawn up did not demonstrate an understanding of the experiments which the Lunar Orbiter Project Office desired to have on board the spacecraft.

Swetnick called a special meeting with Boeing representatives on October 29 for a detailed discussion of Boeing's approach to the experiments. He and Taback made clear to the contractor that Boeing's specifications document for the radiation experiment was very confusing because "it did not in any way provide the bidders with a description of the requirements for the radiation data, a statement of objectives, and a description of what should be done."[34] Boeing's lack of knowledge about the radiation experiment surprised the two NASA officials, who urged Boeing to work out a more realistic approach to fabrication and testing of the experiment's instrumentation as Dr. Foelsche had designed it.

The October 29 meeting revealed the existence of poor communications between Langley and Boeing in the area of experiments. Boeing did not lack the ability to carry out

---

[34] Martin J. Swetnick, Lunar Orbiter Program Scientist, Report on Trip to Boeing on October 27-29, 1964, report dated November 5, 1964, p. 2.

the work required or to obtain competent support for the work. Instead Boeing personnel responsible for the experiments had not understood precisely what Langley desired them to do. Boeing management officials realized that they needed to modify the specifications document to give their bidders a much clearer idea of the nature and objectives of the two experiments. They assured Taback that they would send the modified document to Langley for review and approval before submitting it to the bidders.

The problem with the micrometeoroid experiment was different. Boeing had made certain design changes on it without notifying the principal investigator, Charles A. Gurtler at Langley. Taback and Swetnick were disturbed that Boeing had decided to locate the micrometeoroid pressure cells on the periphery of the tank deck (middle deck) outside the thermal blanket, necessitating reduction of the number of cells from 20 to 15. Worse yet, the leads from the cells to the respective electronics would have to pass through the thermal blanket. Taback made it clear that Langley would have to examine this alteration very carefully before making a decision on the experiment's final design.[35]

---

[35] Ibid., p.1.

Swetnick told the Boeing people that Gurtler did not believe that the experiment could be useful with fewer than 20 cells and any change in their location would require substantial redesign. Again the fact that Langley officials were unaware of Boeing's thinking on the micrometeoroid experiment showed a surprising lack of communication, and steps were taken to strengthen ties between the Langley LOPO people and their Boeing counterparts.

Another problem of note was the status of the Lockheed Agena D launch vehicle, its adapter, and the spacecraft shroud. The Lewis Research Center near Cleveland, Ohio, had the responsibility for these pieces of hardware. Early in 1964 Lewis had insisted that Lockheed handle the entire integration of the booster-adapter-shroud hardware for Lunar Orbiter. Langley had proposed to have Boeing provide the adapter and the shroud. This arrangement had not been acceptable to Lewis. Dr. Abe Silverstein, the center's director, had personally guaranteed that the adapter and the shroud would be delivered to the Boeing Company at the time stipulated in the contract.[36] By late 1964 Lewis was confronted with the predicament that Lockheed, as sole vendor of the hardware, was not going to

---

[36] Report of the LRC and LeRC Lunar Orbiter Shroud and Adapter Meeting, January 5, 1965, p.1. See also Lewis Research Center News Release 65-2, January 6, 1965.

meet the target dates for delivery. Moreover, to meet its schedule might cause it to overrun the original contract price by as much as 100%. Realizing this, Lewis desired to open the field to competitive bidding for the hardware, but it had to wait for a Headquarters review of the situation before making such a move.[37]

Scherer's office at NASA Headquarters was disturbed by the unforeseen turn of events at Lewis. Lockheed had failed to provide Boeing with an adapter master gauge on December 1, 1964, as it had promised; and Boeing still did not have one by January 5. Worse yet Lewis had not finalized the adapter design by the beginning of 1965, and this would impinge upon program schedules unless NASA Headquarters quickly altered the situation. Boeing, meanwhile, had sent Lockheed a model of the spacecraft on January 4 for separation tests with the Agena, but it remained uncrated pending a decision by NASA to open the field for competitive bids for the adapter and the shroud.[38]

By February 8, 1965, Lewis had opened bidding for the spacecraft adapter, the Atlas SLV-3 and the Agena D launch vehicles. Headquarters gave Lewis permission to open

---

[37] Ibid., p.2.
[38] Ibid.

bidding on the shroud, and the bidding began on February 5.[39] On March 8 Lewis awarded Lockheed the adapter hardware contract, and in the interim Lewis delivered the Adapter Master Gauge to Boeing.[40] Boeing, intent upon avoiding any delays or compatibility problems, bid for the spacecraft shroud and was awarded the contract by Lewis on April 1. Boeing would build two ground-test shrouds and five flight shrouds for its Lunar Orbiter.[41] On April 26 Lewis sent Boeing a shroud from the Mariner D spacecraft to be used as a "stand-in" for tests with component sets A and C.[42] These progressive actions by Lewis corrected a situation which could have caused substantial schedule slippage, possibly affecting the incentives in the Boeing contract.

From February 24 through 26, Langley held the Third Quarterly Review. During the review three meetings convened to examine the status of the spacecraft, the results of the Critical Design Review and the interrelations of the

---

[39] Lunar Orbiter Project Office, Langley Research Center, Project Lunar Orbiter, Narrative Analysis, February 8, 1965.

[40] Ibid., March 17, 1965.

[41] Ibid., April 16, 1965.

[42] Ibid., April 28, 1965.

program's various systems: spacecraft, launch vehicle, and tracking and data acquisition.

Boeing reported that the late availability of hardware from Eastman Kodak and RCA had necessitated a schedule adjustment moving prototype systems tests back eight weeks. Beginning in November 1964 Eastman Kodak had to rearrange its schedules with Boeing because its hardware deliveries would not come in time to undergo testing with the spacecraft component sets. Instead Boeing had to use a photographic subsystem simulator during the design verification tests.[43]

By late January 1965 the photo subsystem was still experiencing delays. Eastman Kodak had problems in procuring high-reliability parts and in a power change for the subsystem. The 610 mm lens was also a problem, because of difficulties in attaining the proper resolution; Kodak, however, succeeded in eliminating the error in the lens formula and proceeded with fabrication.[44] The delays did not change the first launch date because the program used the parallel testing mode. However, Langley deleted the Flight Acceptance Test on Spacecraft 1 and established

---

[43] Ibid., December 9, 1964.

[44] Ibid., January 25, 1965.

testing restraints to fit the schedule changes because of the delays at Eastman Kodak.[45]

Boeing also reported to the members of the Third Quarterly Review that all designing was essentially completed and a substantial amount of structural and thermal testing of components had been conducted. No serious failures or deficiencies in components had been uncovered during testing. Nevertheless a few hardware items did have problems: 1) the design and operation of the camera thermal door; 2) telemetry data handling during testing; 3) the photographic recording equipment at DSIF Site 71 (located at Cape Kennedy), and 4) several potential trouble areas in the spacecraft's film processing system. Work on these items did not threaten schedules or hinder the progress of other subsystems in any substantial way, largely because of the loose integration of all subsystems in the spacecraft system design.

Boeing officials also noted at the review that the situation at Lewis was improving and being monitored by NASA Headquarters. Finally, the men present at the Third Quarterly Review decided to have Boeing conduct "qualification tests on S/C 1, one mission simulation test on S/C 2, and

---

[45] Third Quarterly Review, February 24-26, 1965, reported March 2, 1965, pp. 1-2.

phase one of the Goldstone Test on S/C 3...prior to the start of FAT on the first flight spacecraft."[46]

By early March Langley had altered the testing program, removing several conservative features in the initial phase of testing to allow for further schedule compression. At the same time restraints were established which required that 1) the qualification and reliability tests of each component for a flight spacecraft had to be completed before the Flight Acceptance Test on the component could begin and that 2) no FAT of an entire flight spacecraft would commence before the completion of qualification tests on Spacecraft 1, of one mission simulation test on Spacecraft 2, and of the first phase of the Goldstone Test on Spacecraft 3.[47] These steps left little room for any major testing failures without causing serious schedule slippages. This was a risk, but one which was calculated, relying on testing procedures at the component level to catch and correct any design or fabrication anomalies before they could reach the subsystem integration level undetected and have a serious impact on the program's timetable.

---

[46] Ibid., p. 2.
[47] OSSA Review -- March 9, 1965, p. 2.

One example of the early detection of such an anomaly had come to light during the February 17 Photographic Subsystem Critical Design Review. Leon Kosofsky, Headquarters Program Engineer, reported to Israel Taback, Langley LOPO Spacecraft Manager, in a memorandum dated March 4 that "the film processor cannot be stopped indefinitely without the risk of losing the mission due to the sticking of the Bimat web to the exposed film."[48]

This condition meant that either the processor or the mission design would have to be altered. At least some of the film would have to be wasted to keep the whole film and the Bimat processing web (film) advancing at a rate sufficient to preclude any sticking.

The Lunar Orbiter Program Office had to know the time the Kodak SO-243 film and the Bimat could safely remain in contact during a non-photographic period. Kosofsky pointed out that, as matters stood, if this time were 3.5 hours or less, then a typical mission such as that envisioned in Bellcomm report TR-65-211-1 (January 25, 1965) would be impossible.[49] If the safe time was between 3.5

---

[48] Memorandum from SL/Engineer, Lunar Orbiter Program, Lunar & Planetary Programs, to Langley Research Center, Attention: Mr. I. Taback, Lunar Orbiter Project Office, March 4, 1965.

[49] D. D. Lloyd and R. F. Fudali, "Lunar Orbiter Mission Planning," Bellcomm TR-65-211-1, January 25, 1965.

and 6.33 hours, waste exposures would be required on every non-photographic orbit of the Moon, because of the forty-minute processing period which could be subtracted from the time requirement of a photographic and a non-photographic orbit combined. Finally, a safe time of 7.5 hours meant that wasted exposures would be required only on alternate orbits during non-photographic periods, while a 10.5 hours safe time would allow two successive orbits during such periods without having to waste film. This problem presented sufficient potential impact upon Lunar Orbiter's mission capabilities to require immediate study of ways to reduce or eliminate film wastage regardless of the final processor safe time.[50]

The amount of time wasted in the readout process by blank pictures presented one of the worst aspects of the film advance problem. As of March 4, 1965, the design of the photographic subsystem precluded any rapid operation of the rewind drive. Unless changed, this problem would severely affect the critical readout process. Kosofsky instructed G. Calvin Broome, Chief of the Photo Subsystem Section of the Langley LOPO, to explore ways of overcoming the necessity to waste film and prolong the readout

---

[50] Memorandum from SL/Engineer, March 4, 1965.

process.[51]

Except for several minor problems the Lunar Orbiter design phase was completed by April 13, 1965; over 80% of the procurement had been started and over 60% of the first sets of components had been delivered to the contractor. Development tests had begun and mission planning for Orbiter was just commencing. The Kent Testing Facility at Boeing in Seattle also neared completion. Boeing would use it for the spacecraft's mission simulation tests. It consisted of a major chamber with a working section 12 meters high by 9 meters in diameter, capable of having its internal pressure pumped down at twice the rate of the planned Lunar Orbiter ascent profile for the mission simulation tests. Other smaller chambers were also part of this testing facility.[52]

By the middle of 1965 the Lunar Orbiter Program was well into its major development phase. The Program Office and the Project Office at Langley had maintained an equilibrium among the many different needs which had to be fulfilled, and among working groups at Langley, Boeing,

---

[51] Ibid., p. 2. See also memorandum from SL/Engineer, Lunar Orbiter Program, to SL/Manager, Lunar Orbiter Program, March 11, 1965.

[52] OSSA Review -- March 9, 1965, p. 1, and OSSA Review -- April 13, 1965.

the Jet Propulsion Laboratory, Lewis, and the major subcontractors. Langley maintained tight control of its funds and the rate of funding required by Boeing as the program moved into the mission planning phase.

CHAPTER VIII

LUNAR ORBITER MISSION OBJECTIVES AND APOLLO REQUIREMENTS

OSSA and OMSF Planning Activities

While Langley and Boeing accelerated the construction and testing phase of the program, the work of designing the Orbiter missions brought the Office of Space Science and Applications and the Office of Manned Space Flight to a long series of plenary meetings and task group assignments. This work greatly assisted Langley in its own mission planning activities.

The Lunar Orbiter Program was well into its third quarter of operations when Dr. George E. Mueller, Associate Administrator for Manned Space Flight, sent a memorandum to Bellcomm, a contractor to his office, requesting answers to two items fundamental to Apollo site selection: 1) Who held the responsibility for lunar site selection and analysis? 2) Who, where, and how were the films and other data generated by the Lunar Orbiter and the Surveyor Program going to be stored?[1]

Mueller's November 3, 1964, memorandum brought a quick response from Bellcomm. It reviewed the status of work related to lunar site analysis and selection. This became

---

[1] Memorandum from Dr. George E. Mueller, Associate Administrator, Office of Manned Space Flight, NASA Headquarters, November 3, 1964.

the basis for the organization of the Surveyor/Orbiter Utilization Committee. On December 23 Bellcomm reported to Mueller's office that Apollo landing site selection was a function of OMSF. It had the responsibility of defining strategies, goals, schedules, and trajectories with OSSA. The report suggested that OMSF form a working group charged with:

    a. Examining the problem of lunar site analysis and selection.

    b. Recommending the initiation of any work necessary.

    c. Making recommendations on any new facilities needed for the adequate analysis and storage of the data.

    d. Examining the necessary funding and identifying the responsible organizations.

    e. Identifying the manner in which landing site selection should be accomplished.[2]

The proposed working group would consist of a chairman reporting either to the Associate Administrator for Manned Space Flight or to the Apollo Program Director, Maj. Gen. Samuel C. Phillips. The Office of Space Science and Applications would assign representatives from the Surveyor and the Lunar Orbiter Programs. The Manned Space Flight Center would assign representatives from the Apollo Spacecraft Project Office, the Flight Operations Division, and the Flight Crew Operations Division. Manned Space Flight Operations and Manned

---

[2] Memorandum from T. H. Thompson, Bellcomm, Inc., to Dr. G. E. Mueller/Gen. S. C. Phillips, December 23, 1964.

Systems Engineering in the Office of Manned Space Flight, with the Bellcomm Site Survey Group, would also appoint representatives. Lastly, the Bellcomm memorandum to Mueller recommended that Myron W. Krueger, the OMSF man responsible for lunar photographic data, be assigned.[3] This would form the nucleus of the more formal Surveyor/Orbiter Utilization Committee which came into being at a later date.

As of December 23, 1964, the Office of Manned Space Flight had no organization to accept and store Surveyor or Lunar Orbiter data. No organized group existed to perform lunar site analysis and selection. The Apollo Project Development Plan stated the need for a working group to make recommendations to the appropriate groups within OMSF on the optimum utilization of data, but no such group had been set up. On the other hand the Lunar Orbiter Project Office had already set up a working group to make recommendations on the form of data and its storage and retrieval. And Bellcomm's Site Survey Group monitored site survey programs for Lunar Orbiter and Surveyor and developed strategies for the use of systems in these programs.[4] The time had come for the Office of Manned Space Flight and the Office of Space Science and Applications to form firmer working relations.

---

[3] Ibid.
[4] Ibid., Attachment A--Review of Current Status of Work Related to Lunar Site Analysis and Selection.

179

On September 22, 1964, Oran W. Nicks had informed the Apollo Program Director, General Phillips, about the mission planning effort that the Lunar Orbiter Program was undertaking at Langley. This effort could possibly influence Apollo hardware design. Nicks suggested that OMSF make a study of specific Lunar Orbiter missions in support of Apollo. The recommendations of the study would aid the Lunar Orbiter Program Office in developing guidelines for actual mission planning activities at the Langley Research Center and at Boeing. Nicks pointed out that Bellcomm had very qualified men to make such a study for OMSF.[5]

Nicks's memorandum resulted in a Bellcomm study for OMSF during the remainder of 1964. On February 18, 1965, Phillips sent Nicks the report of the study, "Lunar Orbiter Mission Planning," by Douglas D. Lloyd and Robert F. Fudali of Bellcomm. Phillips expressed a willingness to have further joint study done if Nicks agreed that it was necessary.[6]

The Lloyd-Fudali report explained that Lunar Orbiter could take nearly identical photographs in different ways.

---

[5] Memorandum from SL/Director, Lunar and Planetary Programs, to MA/Maj. Gen. Phillips, Office of Manned Space Flight, September 22, 1964.

[6] Memorandum from MA/Apollo Program Director to SL/Lunar and Planetary Programs Director, February 18, 1965.

Two simulated missions were described in the report, one in a posigrade orbit, the other in a retrograde orbit. Further, the study had reached the following conclusions:

1. The strategy of contiguous high-resolution photography of multiple targets should be used. This would permit successful site survey with only a single Lunar Orbiter.

2. To allow the above, the camera sequencer control should be changed to include a quantity control for providing eight consecutive photographs.

3. The quantity of gas made available for the attitude control system should be sufficient for a minimum of sixteen separate photographic manuevers.

4. To achieve at least 1-meter optical pair resolution, photographs should be taken from a nominal height of 46 km or less.

5. To avoid the possible problem of orbital instability for the above low-altitude orbit, because of the uncertainties in knowledge of the moon's spherical harmonic terms, the orbit should be inclined no more than 7° to the lunar equator.[7]

Further Bellcomm research during March 1965 produced a paper entitled "Apollo Lunar Site Analysis and Selection," which was transmitted to General Phillips. Pointing out that Lunar Orbiter and Surveyor were the two prime data-gathering systems for Apollo, it recommended that OMSF and OSSA set up a Joint Site Survey Steering Committee. Its major task

---

[7] "Lunar Orbiter Mission Planning," Bellcomm, Inc., January 25, 1965, p. ii.

would be the definition of the objectives and use of Lunar Orbiter and Surveyor for the Apollo Program's needs. The committee would have the responsibility for target selection, launch schedules, choice of measurements, measurement priority and instrument complement, control of data handling, and recommendations on data analysis for each Lunar Orbiter and Surveyor mission.[8]

On May 10 Brian T. Howard of Bellcomm reported to General Phillips that, in addition to earlier recommendations for Lunar Orbiter and Surveyor tasks in Apollo site selection, Bellcomm had considered two more proposals related to the organization of cooperative OMSF-OSSA activities in site analysis and selection. First, it seemed highly desirable to set up a joint OMSF-OSSA Lunar Surface Working Group. It would report to the Apollo Program Office and to the Lunar and Planetary Programs Office. It would coordinate mutual planning activities concerning site survey requirements and the ways in which they could be satisfied. Second, Bellcomm recommended that the Manned Space Flight Center's Data Analysis Division subcontract with JPL for the prime responsibility of gathering, analyzing, and evaluating data.[9]

---

[8] "Apollo Lunar Site Analysis and Selection," Bellcomm, Inc., March 30, 1965.

[9] Memorandum from B. T. Howard, Bellcomm, to Maj. Gen. S. C. Phillips, NASA/MA, May 10, 1965.

Developing Mission Designs

While Bellcomm was advising OMSF, the Langley Lunar Orbiter Project Office carefully studied and compared the proposed missions that Bellcomm had developed (i.e., in the Lloyd-Fudali report) with the one developed by Boeing. Thomas Young of the Langley LOPO informed Norman L. Crabill on May 7 of the conclusions pertaining to the reliability of each proposed mission. His memorandum stressed the differences in reliability in the studies performed by Bellcomm and Boeing. The Bellcomm mission required 4.5 days longer to accomplish than did that of Boeing, but the variation in resulting data was minimal.[10]

Young's LOPO mission planning study group continued to analyze Lunar Orbiter capabilities and concluded in a report to Crabill on June 14 that Apollo and Surveyor requirements permitted variable Lunar Orbiter missions, ranging from a concentrated to a distributed photographic mission, depending upon primary requirements for the two programs. For photographic missions with sites distributed within the Apollo zone, a set of trajectories could be defined that were generally independent of the exact locations of the sites. They could be planned by placing mild

---

[10] Memorandum from A. T. Young to N. L. Crabill, Langley Research Center, May 7, 1965, Subject: Mission Reliability Analyses and Comparison for the Bellcomm Mission and TBC's S-110 Mission.

restrictions on the latitude range of the sites. Thus, for Missions I, II, and III (with prime sites in the Apollo zone), trajectories could be defined without consideration of the exact site locations. Mission II sites were to be selected from the review of the results of secondary sites of Mission I, and Mission III sites were selected from all results of the first two missions.[11] However, the Langley Project Office considered the establishment of mission objectives a prerequisite to further mission planning.[12]

On Friday, June 25, representatives from OSSA, OMSF, the Langley Lunar Orbiter Project Office, the Manned Space Flight Center, the Jet Propulsion Laboratory, and Bellcomm held the initial coordination meeting to establish a preliminary plan for utilizing Lunar Orbiter's mission capabilities with the first Lunar Orbiter mission, the first Surveyor mission, and with Apollo mission requirements. During the meeting it was agreed that the Lunar Orbiter could best aid Surveyor by screening sites and defining targets which had a high probability of being smooth. The

---

[11] Memorandum from Norman L. Crabill, Mission Analysis and Design Engineer, Viking Project Office, Langley Research Center, to NASA Code EH, Attention: Dr. Eugene M. Emme, December 9, 1969.

[12] Memorandum from A.T. Young to N.L. Crabill, Langley Research Center, June 14, 1965, Subject: Lunar Orbiter Mission Planning Study, pp. 1, 6.

representatives from the Apollo Systems Engineering Office stated that Lunar Orbiter could photograph a landed Surveyor spacecraft from an altitude of 46 kilometers with 1-meter resolution because of the Surveyor's shadow at a prescribed Sun angle and the high albedo of the spacecraft. Lunar Orbiter had originally been targeted to screen Surveyor sites. After a Surveyor had successfully landed, the Orbiter was to overfly it and photograph it through the 610 mm high-resolution camera lens. The increased capabilities of the Lunar Orbiter photo subsystem now allowed it to combine screening and overfly tasks in the high-resolution mode.[13]

The Apollo Systems Engineering Office and the Manned Space Flight Center preferred that Lunar Orbiter fly a distributed mission; this offered a sampling technique better able to find an area suitable for an Apollo landing, to define suitable areas for further coverage on later Orbiter flights, and to increase the flexibility of the Apollo launch window by finding suitable sites spread across the Apollo zone of interest. Both the Manned Space Flight Center and Bellcomm recommended that Lunar Orbiter photograph the Ranger VIII impact point located in the Apollo zone because possibly it could serve as a future

---

[13] Minutes: Lunar Orbiter Target Objectives Meeting at Langley Research Center, June 25, 1965, recorded by A. Thomas Young, pp. 2-3.

Apollo orbit anchor point.[14]

The June 25 Langley meeting provided the Lunar Orbiter Project Office with information concerning mission objectives from the Apollo and the Surveyor Program Offices. This assisted Langley in its mission planning activities, and it, in turn, was better able to guide the Boeing Company in its work.[15] Moreover, the meeting produced the basis for efficient coordination between the NASA offices requiring Lunar Orbiter data and enabled the Lunar Orbiter Program to develop preliminary mission plans.[16]

From July 13 to 15 a preliminary mission definition meeting for Lunar Orbiter convened at Langley. The men present[17] defined preliminary mission types on the basis of decisions arising out of the June 25 meeting at Langley. These mission types depended upon three basic flight objectives: 1) gathering significant topographic information of the Moon's surface for selection of Surveyor and Apollo

---

[14] Ibid., pp. 4-6.

[15] Memorandum for File, from Dennis B. James, Bellcomm, Inc., June 30, 1965, Subject: Trip Report: Lunar Orbiter Mission Planning Meeting -- Langley Research Center -- June 25, 1965.

[16] OSSA Review -- July 2, 1965, p. 3.

[17] Attendees were: D. D. Viele, Boeing; Douglas D. Lloyd, Bellcomm Leon J. Kosofsky, NASA Lunar Orbiter Program Office; Clifford H. Nelson, Norman L. Crabill, Gerald W. Brewer, and A. Thomas Young, Lunar Orbiter Project Office, Langley.

sites; 2) providing selenodetic data on the size, shape, and gravitational properties of the Moon necessary for determining orbit lifetime of a Lunar Orbiter sufficiently long to allow adequate time for readout; and 3) providing measurements of micrometeoroid and radiation flux in the lunar environment.[18]

By the end of July the Lunar Orbiter Program Office in Washington had the results of the Langley LOPO and Bellcomm preliminary mission studies. Four mission types had been formulated on the basis of requirements and recommendations from Apollo, Surveyor, and Lunar Orbiter Program Offices. Briefly summarized they were:

> Type I --Site sampling, a distributed mission allowing eleven single passes over different terrains (i.e., highlands, maria, rilles).
>
> Type II --wide-area coverage for Surveyor of only three separate sites.
>
> Type III --Surveyor location mission to pinpoint landed Surveyor at one-meter resolution.
>
> Type IV --a combination mission for more sophisticated work later in the program.[19]

A joint OSSA/OMSF Site Survey Meeting was held at NASA Headquarters on August 4 to review the status of the Surveyor, Lunar Orbiter, and Apollo Programs and to discuss

---

[18] N. L. Crabill and A. T. Young, "Preliminary Lunar Orbiter Mission Types," Lunar Project Office, July 16, 1965, p. 1.

[19] OSSA Review -- July 30, 1965, pp. 2-3. See also Crabill and Young, "Preliminary Lunar Orbiter Mission Types."

preliminary mission planning for Lunar Orbiter and selection of Surveyor landing sites. Clifford H. Nelson, Lunar Orbiter Project Manager, summarized the status of the Lunar Orbiter Program and pointed out that the program expected to meet its original launch schedule but that slips in subsystems, especially the photographic subsystem, had necessitated further compression of the testing schedule in order to hold the launch schedule.[20]

After Nelson's report and the Apollo status report, Norman L. Crabill presented the preliminary planning for the first two Lunar Orbiter mission types. He outlined the ground rules for the Type I mission:

Ground Rules

    1) Photograph two sites of each smooth-looking-terrain class up to a total of eleven sites within the Apollo area of interest.

    2) Photograph Ranger VIII and any landed Surveyors.

    3) Photograph each site using a single pass with sixteen contiguous 1-meter-resolution frames per pass.

    4) Read out up to four frames between passes.

    5) Define mission for the Boeing Company by the fall of 1965.

And for the Type II mission:

Objectives

    1) Topography mapping for possible Surveyor sites.

---

[20] SSA/MSF Site Survey Meeting, Minutes, August 4, 1965, document dated August 12, 1965, Bellcomm File, pp. 3-4.

2) High-precision selenodetic data.

3) Lunar environmental data.

Ground Rules

1) Photograph three sites spread 30° of longitude apart.

2) Use four passes per site.

3) Use sixteen high-resolution contiguous frames per pass.[21]

At the August 4 meeting Lee R. Scherer proposed the establishment of a Lunar Photographic Analysis Steering Group which would act as a sounding board for suggestions and requests from the various programs involved in lunar exploration. It would also establish priorities and serve as coordinator for NASA-wide activities related to obtaining photographic data of the Moon. The group could coordinate such activities as control of Earth-based lunar mapping, direction and planning in the analysis of Lunar Orbiter data, monitoring of pertinent work for other government agencies, planning with the OSSA planetology group, handling agreements for data processing priorities, and coordinating Apollo needs with other requirements. No final action was taken on Scherer's proposal at the meeting, but it stimulated discussion on these aspects of mission planning and data utilization.[22]

---

[21] Ibid., pp. 5-6.
[22] Ibid., p. 8.

The Ad Hoc Surveyor/Orbiter Utilization Committee (SOUC)

All of the previously discussed plenary meetings served as the basis for setting up the OSSA/OMSF Ad Hoc Surveyor/Orbiter Utilization Committee, which held its first meeting on August 20, 1965.[23] At this time Scherer reviewed the Lunar Orbiter photographic format and described the photographic subsystem in detail. Following this he stressed these major points which had to be considered in Orbiter mission planning:

1) Resolution and area coverage are directly proportional to orbital altitude.

2) A photographic pass requires an altitude manuever.

3) The system can take 1, 4, 8, or 16 pictures on a single pass.

4) The system is capable of taking 192 pictures total.

5) The last 4 pictures in the take-up spool can be read out on command anytime during the mission.

6) The system is capable of reading out one frame during each orbit. Pictures cannot be taken during the readout.

7) The thread-up distance from the camera to the readout is 18 frames.

8) Total readout will be accomplished after completion of all photography; the last photograph taken will be the first read out.

---

[23] Members of the Surveyor/Orbiter Utilization Committee were: Edgar M. Cortright (Chairman), OSSA; Samuel C. Phillips (Apollo Program Office), OMSF; Edward E. Christensen (Manned Operations), OMSF; William A. Lee (ASPO), OMSF; William E. Stoney (Data Analysis), MSC; Oran W. Nicks (Lunar and Planetary Programs), OSSA; Urner Liddel (Lunar and Planetary Science), OSSA; Lee R. Scherer (Lunar Orbiter Program), OSSA; Benjamin Milwitzky (Surveyor Program), OSSA; Victor Clarke (Surveyor Project), JPL; Israel Taback (Lunar Orbiter Project), Langley.

9) Gravity perturbations and latitude width of good lighting both increase with orbital inclination. There will have to be some trade-off studies made in this area; what's good for selenodesy doesn't produce the best pictures.[24]

Norman L. Crabill followed Scherer with an updated outline of the four mission types which Langley had developed for Lunar Orbiter:

> Type I -- Photographs ten evenly distributed target sites in the Apollo zone of interest and covers each site in high- and low-resolution stereo photography (1 meter and 8 meters).
>
> Type II -- Photographs four sites to screen for Surveyor landing sites in Apollo zone.
>
> Type III -- Photographs to 1-meter resolution an area containing a landed Surveyor to learn as much as possible about the surrounding terrain.
>
> Type IV -- Obtains a variety of topographic data not obtained by other mission types.[25]

The ordering of these mission types reflected the conservative philosophy of OSSA and Langley covering the Lunar Orbiter mission objectives. It was vital to obtain reliable, accurate data for the Apollo Program before attempting to do anything else. Thus the first mission type was entirely devoted to Apollo's needs. Also, the mission planners had to take into consideration the

---

[24] Ad Hoc Surveyor/Orbiter Utilization Committee Minutes First Meeting, Washington, D.C., August 20, 1965, pp. 2-3.

[25] Ibid., pp. 4-5.

possibility of a spacecraft or mission failure, in which case they wanted to have as many remaining Orbiters to carry out the Apollo photographic reconnaissance mission as possible. Were the Lunar Orbiter Program strictly pursuing scientific objectives unrelated to Apollo, a general survey mission of the entire Moon from a high polar orbit would have been preferable as the first mission. This was not the case.[26]

The SOUC agreed to let Scherer define the decisions and the dates for the next meeting. The Committee requested him to tell Boeing to concentrate on studies of multiple and distributed targets instead of studying models for large block photography of the Moon's surface. The Committee also asked Scherer to hold a working meeting of representatives from the Apollo, Surveyor, and Lunar Orbiter Programs to determine the preliminary plan for the first Lunar Orbiter mission. The Committee favored a distributed Type I mission and asked that a presentation of the first mission plan be made within thirty to forty-five days.[27]

The prime role in mission planning was carried out by

---

[26] Recorded Interview with Israel Taback, former Lunar Orbiter Spacecraft Manager, Langley Research Center, July 7, 1970.

[27] Ad Hoc Surveyor/Orbiter Utilization Committee Minutes...August 20, 1965, p. 1.

the Langley Research Center while the SOUC acted in an advisory way, coordinating activities among the various centers connected with the Lunar Orbiter Program. The working meeting requested by SOUC took place at Langley on September 8 and 9. It had the following major objectives:

> 1) To gain understanding of Orbiter and Surveyor mission design problems.
>
> 2) To review available data on the lunar surface.
>
> 3) To produce lists of lunar sites which would satisfy Apollo, Surveyor, and Lunar Orbiter constraints.[28]

At the meeting Scherer pointed out that Homer E. Newell, NASA Associate Administrator for Space Science and Applications, would have to make the final decision on the first mission plan for Lunar Orbiter and that he would rely on recommendations from Langley and SOUC. Therefore, the Lunar Orbiter Program Office would be required to present a detailed, well-defined plan to the Surveyor/Orbiter Utilization Committee.[29]

The Apollo Spacecraft Program Office (ASPO), represented by James Sasser from the Manned Space Flight Center, Houston, Texas, expressed its desire for a Lunar Orbiter distributed mission and concurred on the sampling of

---

[28] Lunar Orbiter Mission Planning Meeting, Langley Research Center, Bldg. 1251, Rm. 105, September 8-9, 1965, Minutes recorded by A. T. Young.

[29] Ibid., p.1.

different terrain types within the Apollo zone of interest with emphasis on the areas of greatest apparent smoothness. However, ASPO did not want the Lunar Orbiter restricted to sampling Surveyor-size landing areas or sites accessible only to the Surveyor spacecraft. As a result Sasser accepted an action item to provide the Lunar Orbiter Project Office with a letter confirming the bounds of the Apollo zone of interest.[30]

Lawrence Rowan of the United States Geological Survey made a presentation to the members of the meeting in which he discussed the USGS lunar terrain analysis based upon the newest lunar map from the Aeronautical Chart and Information Center (ACIC) with a scale of 1:1,000,000. Rowan talked about the various sources of data that went into making the lunar map and then gave an interpretation of terrain types on the Moon. The USGS terrain analysis enabled Rowan to present a list of nine terrain types to be sampled photographically by Lunar Orbiter: 1) dark mare, 2) mare, 3) mare ridges, 4) mare rays, 5) upland Unit-I, 6) deformed crater floors, 7) upland Unit-II, 8) crater rims, and 9) sculptured highlands.[31] Rowan's information formed part of the basis for the site selection process which followed.

---

[30] Ibid., p.3.
[31] Ibid., pp.3-4.

The members of the meeting subsequently developed two Orbiter missions based upon the USGS terrain map and the following assumptions: 1) orbital inclination of spacecraft equals 12.5°, 2) descending-node photography to be employed, 3) orbital spacing to be based on Goudas's model of the Moon, 4) lighting band to be based on a spherical Moon, and 5) lighting band to be initially centered about the lunar equator at 0° longitude.[32]

Two preliminary mission plans resulted. Members at the meeting subsequently picked them apart and criticized various aspects. Their major criticism was that the plans included too many samples of mare terrain types. They generally agreed that on the first mission Lunar Orbiter should photograph only the Apollo zone of interest unless a Surveyor landed outside of it.[33] The results of the Langley meeting formed the foundation of the Lunar Orbiter Mission A plan.

Presentation of Mission A

On September 29, 1965, the Lunar Orbiter Project Office at Langley formally presented the Mission A plan to the Surveyor/Orbiter Utilization Committee. It would be a

---

[32] Ibid., pp. 4-7.
[33] Ibid.

Type I mission, sampling various lunar surface areas in the Apollo zone of interest. Lunar Orbiter's camera would assess selected sites for their suitability for Apollo and Surveyor landings.[34] An excerpt from the OSSA Review briefly describes Mission A:

> A few pictures will be taken on the initial orbit. The location could range from 60° east to 110° east and will be determined later. In the final orbit, ten separate sites will each be covered by a single photographic pass. Briefly, site one is the only example of a dark mare in the Apollo areas of interest. Dark mare are considered the smoothest of the various terrain types. Site two is a highland site with smooth basins. Site three is in the same longitude as Ranger VIII. It is a ray mare probably not quite as rough as shown by Ranger photographs. Site four is a highland site which will contain photographs of each of the four highland terrain units. Site five, in Sinus Medii, has high potentiality for Apollo and Surveyor landing areas. Site six contains upland units and a deformed crater floor. Site seven is a good example of a mare with sinuous ridges. Site eight is a smoother mare with linear ridges. Site nine is located in the old crater floor Flamsteed and is probably the prime Surveyor landing site at this time. Site ten is outside of the Apollo area but is a dark mare and may be utilized for Surveyor.[35]

Langley had done a thorough job of screening each area for compatibility with Apollo and Surveyor needs and with Lunar Orbiter photographic capability. The Committee approved the plan.

---

[34] Lunar Orbiter Project Office Recommendation for Lunar Orbiter Mission A, presented to the Ad Hoc Surveyor/Orbiter Utilization Committee, September 29, 1965.

[35] OSSA Review--October 5, 1965, p.1.

After winning the SOUC's approval for Mission A Scherer made a presentation to a meeting of the Planetology Subcommittee of the OSSA Space Science Steering Committee on October 21 and 22. With him were Harold Masursky and Lawrence Rowan of USGS. Scherer reviewed the procedure for selecting the ten areas on the lunar surface which the first Lunar Orbiter would photograph. He stressed that the mission's objective was to obtain detailed topographic data for assessing the suitability of specific areas as possible Apollo and Surveyor landing sites.[36]

Masursky explained in detail how the Lunar Orbiter Program could apply the methods of structural and stratigraphic geological mapping developed for Earth studies when these were augmented by telescopic observations and the Ranger pictures of the Moon. Rowan outlined recent findings concerning crater densities, surface roughness, and albedo of the Moon. He specifically described the ten selected areas which Lunar Orbiter would photograph on Mission A. He also stressed that the USGS work had led him to conclude that crater density measurements were not too useful in the selection of landing sites, but they aided in distinguishing between rayed and non-rayed surfaces. This, he pointed out, suggested a relationship between surface

---

[36] Summary Minutes: Planetology Subcommittee of the Space Science Steering Committee, October 21-22, 1965, p.8.

roughness and albedo.[37]

Following this meeting the Planetology Subcommittee drew up a resolution, based upon the Lunar Orbiter Program Office's reports and the USGS information, which it forwarded to Oran W. Nicks. Although the resolution did not influence mission plans for the first Orbiter, it showed the Subcommittee's direction of thinking:

> The Planetology Subcommittee is disturbed that there are no scientific missions planned to take advantage of the unique capabilities of Lunar Orbiter for conducting investigations of the Moon, after the five flights in support of Apollo and Surveyor lunar landing site selection. In view of the opportunity to perform certain experiments (geodesy, gamma ray, x-ray, magnetometry, microwave, and non-imaging radar) in orbit about the Moon before the Apollo Applications Program, the Subcommittee recommends that every effort be made to undertake Lunar Orbiter scientific missions at the earliest possible date.[38]

The Subcommittee did recognize the priorities which placed Apollo and Surveyor requirements before any purely scientific objectives in the Lunar Orbiter Program and at its Spring 1966 meeting recommended "that major attention be given to photography of sites of scientific interest, following the initial, successful Lunar Orbiter flight. These data are of particular importance in the planning and

---

[37] Ibid., pp.8-9.

[38] Memorandum from SL/Chairman, Planetology Subcommittee (Dr. Urner Liddel), to SL/Director, Lunar and Planetary Programs, Subject: Resolution on Lunar Orbiter Scientific Missions, November 5, 1965.

ultimate scientific value of both manned and unmanned lunar surface missions."[39]

Mission planning activities continued to develop Lunar Orbiter's role in fulfilling Apollo and Surveyor requirements during the remainder of 1965 and the first quarter of 1966. Funding and hardware problems in the program made up the other significant activity during 1965.

Funding and Technical Problems--1965

During the course of 1965, funding and technical problems exerted significant influence upon the Lunar Orbiter Program's schedules. Already in April 1965 the total projected cost of the program was up by $10 million, of which $4.5 million was required in fiscal 1965. Scherer expressed surprise at this increase because NASA had been maintaining very close communications with Boeing.[40]

Langley had known early in February that the total estimated cost of the Boeing contract was about $94.8 million, of which $4 million was to be spent for authorized changes and $10.3 million for estimated overruns.[41] By

---

[39] Planetology Subcommittee of the Space Science Steering Committee, Meeting No. 4-66, May 9-11, 1966, p.16.
[40] OSSA Review--May 6, 1965, p. 1.
[41] Project Lunar Orbiter, Narrative Analysis, Langley Research Center, February 8, 1965.

mid-March the cost picture had changed slightly: $96.4 million for the Boeing contract, $4.4 million for authorized changes, and $11.5 million for estimated overruns.[42] By the end of March Langley had changes under review amounting to $7.9 million which were not yet authorized.[43] The situation did not seem to reach a plateau and level off, and on April 26 Langley and Boeing began discussions to curb rising costs and keep expenditures within planned funding levels.[44]

One problem in the funding situation had arisen in communications between Boeing and the two major subcontractors: Eastman Kodak and RCA. The majority of the overruns were occurring in their operations. Eastman Kodak projected an increase of 26% in costs and RCA a 32% increase over original estimates. The estimates reflected a basic underestimation by Boeing management of the costs of the hardware the two subcontractors were obligated to supply. Boeing had had inadequate communications with the two companies during contract negotiations, and the talks had taken an unusually long time to reach final agreements. Langley realized that the situation could be controlled only

---

[42] Ibid., March 17, 1965.

[43] Ibid., March 31, 1965.

[44] Ibid., April 28, 1965.

through vigorous cost reduction efforts among all participants in the program. As things stood, the program had $49.5 million for FY 1965, which meant that $5.8 million in unfilled orders would carry over into FY 1966.[45] Boeing also realized that in order to protect its incentives in the contract, it would have to make an effort to reduce the pace of expenditures while tightening up schedules with Eastman Kodak and RCA.

NASA Headquarters directed Langley to conduct specific cost reduction studies to combat surprise jumps in the expenditure rate. Langley requested the same of Boeing. Both actions were initiated at the beginning of May. By May 4 the Lunar Orbiter Project Office had turned up 32 items where potential cost reduction might be possible. At the same time Langley and Boeing officials visited Eastman Kodak and RCA. Their purpose was to bring under control the costs of these two subcontractors, to prevent surprises such as the $10-million jump which had occurred in April, and to submit recommendations for cost saving items which would not affect schedules or disturb performance incentives.

Boeing officials conferred with Langley on May 11 and 12. They informed Langley that Boeing was assigning one

---

[45] OSSA Review--May 6, 1965, p. 2.

assistant project manager to RCA and one to Eastman Kodak. These two officials would control changes in negotiations for changes and keep completely informed of cost projections. Moreover, Boeing would send Langley and NASA Headquarters weekly cost project statements. The assistant project managers assigned to RCA and Eastman Kodak were answerable directly to Robert J. Helberg, the Boeing Lunar Orbiter Program Manager.[46]

In addition to strengthening its management Boeing submitted 53 specific items for cost reduction consideration. Nelson and Scherer were pleased at the rapidity and extent of the Boeing probe for ways to cut costs. The 53 items totaled approximately $8.8 million, of which, by June, NASA had accepted over $4 million. There was still $1 million in items being reviewed for possible cost reduction.

Some specific examples of major items deleted or reduced were: 1) The program ended the requirement to use the RCA test chamber as a back-up for the Boeing chamber at the new Kent facility in the testing phase, saving $280,000. 2) The need for, and frequency of, certain kinds of documentation was reduced, saving $40,000. 3) The redundancy of photo-receiving equipment at the Deep

---

[46]OSSA Review--June 7, 1965, p. 1.

Space Instrumentation Facility sites was reduced, saving $250,000. 4) The need to perform burn-in on all electronic parts of the photographic subsystem at Eastman Kodak was altered to encompass burn-in of certain selected parts where this process had merit, further saving $350,000.[47]

Boeing and Langley program representatives met at Langley on May 11 to discuss cost reductions. Langley decided that because of funding problems in FY 1965 it would fund Boeing on the basis of actual costs for the remainder of the fiscal year which ended on June 30.[48] By the third week in June Langley and the contractor had reached agreement on 22 specific items for cost reduction at an estimated savings of $4 million. Other items were undergoing further cost reduction review.[49]

The decision to reduce by one the number of test spacecraft was a major change in the development phase. While it was part of the cost reduction efforts, this change increased the risk of an operational failure. As originally planned, Set C of the components was to be built up into subassemblies for system testing. After this use, it was to become a complete spacecraft for system design verification

---

[47] Ibid., pp.1-2.
[48] *Project Lunar Orbiter, Narrative Analysis*, Langley Research Center, May 12, 1965.
[49] Ibid., June 23, 1965.

(SDV). Qualification testing was to be performed with Spacecraft 1. Spacecraft 2 was to be used for mission simulation tests, and Spacecraft 3 was scheduled for performance tests at the Goldstone DSIF site and for integration tests at the Eastern Test Range at Cape Kennedy. The change would have the last two tests performed with the spacecraft built from the Set C components. Spacecraft 3 would be assembled according to the existing schedule. It would become a flight spacecraft unless required for further testing. Should it be required for either of the last two tests, it would, nevertheless, be refurbished and used later as a flight spacecraft. Boeing agreed to this, making it possible to build one less spacecraft at a saving of $1.8 million.[50]

Lunar Orbiter Program Manager Scherer felt that the entire cost reduction effort of April, May, and June had proved valuable for the program. The schedule was very tight and events in the program were moving faster. This effort had forced people to re-evaluate themselves, their procedures, and the requirements of their jobs, and it had generated a new respect for cost effectiveness. Exactly how much would be saved in the long run was unpredictable, but Scherer believed that the impact of the cost reduction effort would certainly increase the likelihood that the

---

[50]OSSA Review--June 7, 1965, pp. 1-2.

program would meet its launch schedule dates and that planning and management would become more effective.

The Quarterly Review of mid-June at the Boeing Company indicated that the program would indeed keep its original launch date schedule. Boeing had brought hardware problems under control, save for the line scan tube which had already caused a three-week schedule slip in the photo subsystem.[51] The photographic subsystem still remained the pacing item of the program. Boeing and NASA were completing required test and storage facilities on schedule while twenty-eight of the thirty-three major Lunar Orbiter components were in their testing programs.

The critical testing phase of the program would tell whether or not the original launch dates could be met. During the summer, while Mission A was being developed, several significant hardware problems arose to hamper progress. The line scan tube of the readout subsystem had been failing tests, but by the end of July a new assembly procedure had eliminated the cause of failure. Excessive heat during the sealing of the glass envelope had been damaging the drum bearing on which the tube rotated, causing the electric motor to stall after a few hours of operation. A new tube was fabricated once the problem had been pinpointed,

---

[51] OSSA Review--July 2, 1965, and July 30, 1965.

and it successfully completed a 200-hour test. This delay affected schedules of the ground spacecraft, but did not alter the flight spacecraft schedules.

The propellent tanks of the velocity control engine also presented a problem. Bursting during pressure storage tests at the Bell Aero Systems Company, they seemed to show significant stress corrosion of the titanium alloy by the oxidizer. This complication necessitated a major meeting among Orbiter, Apollo, and Bell officials at North American, the prime contractor for Apollo, to review the history of the tanks. The Apollo Program, the prime user of these tanks, would have to find the reason for failure before Lunar Orbiter Program officials could accept the tanks for use in their spacecraft. In the meantime Boeing decided to use boiler plate oxidizer tanks whenever possible during the testing program to avoid further delays.[52]

By September 9 Boeing was conducting its own testing program of the Bell tanks, subjecting ten of them to tests in various configurations to determine their safety margin for Orbiter applications. OSSA also requested NASA's Office of Advanced Research and Technology to perform basic research to define the specific phenomenon causing the tanks to burst.

---

[52] *Project Lunar Orbiter, Narrative Analysis*, Langley Research Center, August 18, 1965.

Despite tests the tanks remained an unresolved problem. The problem could not be pinpointed quickly, and early in November the Lunar Orbiter Program Office reluctantly decided to decrease stress levels by installing heavier, thicker-walled tanks with a weight penalty of two kilograms.[53] Fortunately this addition did not absorb the remaining weight margin for the spacecraft, which was relatively generous by design.

A problem of leakage in the nitrogen tank was more easily overcome during the same period. Nitrogen, a gas of low atomic weight, was detected leaking through teflon bladders and saturating the oxidizer for the velocity control engine. The bladders were subsequently coated with a layer of aluminized mylar which eliminated leakage.[54]

Progress was also hindered when Boeing Lunar Orbiter personnel discovered excess drift in the inertial reference unit (IRU) of one of the ground spacecraft. An investigation revealed dirty gyros. The discovery necessitated examination of all gyros for the IRUs in the remaining spacecraft, a task which would hold up completion of the attitude control subsystem by thirty days. Boeing disassembled nine of twenty-

---

[53] OSSA Review--September 9, 1965, pp. 1-2, and November 2, 1965, p. 2.

[54] Costello interview, July 9, 1970.

nine gyros that Sperry Rand, the fabricator, had delivered. All nine were found to be badly contaminated.[55] By the beginning of November Sperry Rand had reworked four of the nine, but this rate was insufficient if an impact on the schedules was to be avoided. Yet the time factor would be doubled if NASA decided to procure gyros from another vendor, a fact which clearly revealed that Boeing and Langley were all but frozen to their present course.[56]

These setbacks had not yet jeopardized the schedules of the flight spacecraft, and overall progress was good. The major exception by November was the delivery of Flight Spacecraft 3. Delays in the delivery of the photographic subsystem had caused slippage in its delivery. By late October Lunar Orbiter management had narrowed the reason behind Eastman Kodak's failure to meet schedules to two hardware items: the shutter for the 60-mm-focal-length lens and the Velocity-over-Height (V/H) sensor. Both of these were being manufactured by a subcontractor to Eastman Kodak, Bolsey Associates, Inc.

Langley sent James S. Martin, the Lunar Orbiter Assistant Project Manager, to talk with Eastman Kodak and

---

[55] Boeing Quarterly Technical Progress Report, July to September, 1965, Section II, p. 17.
[56] OSSA Review--November 2, 1965, p. 2.

Bolsey officials about schedules. Martin found that although Eastman Kodak and Bolsey had very qualified people performing the work for Lunar Orbiter, their management did not seem to place great significance on meeting schedules. Bolsey, a small firm of about 80 people, had only the V/H sensor and the focal plane shutter as its two major jobs on a cost-plus-fixed-fee contract. The company had absolutely no financial incentive to accomplish its work on time. Bolsey's work affected the work at Eastman Kodak, which in turn impacted upon the delivery date of Spacecraft 3.[57]

Martin insisted on major corrective actions in coordination and control by Boeing and Eastman Kodak management. Subsequently, Eastman Kodak assigned six full-time persons to the Bolsey plant. The Lunar Orbiter Project Office at Langley followed up Martin's initial visit with a complete schedule review on November 5 and followed this with another visit to Bolsey on November 10.[58] Martin's investigations revealed that each firm had the technical competence to do the work, but neither was particularly devoted to completing its work within the given time. This situation caused extensive delays, permitting the photographic

---

[57] Martin interview, July 7, 1970.

[58] OSSA Review--November 2, 1965, and Project Lunar Orbiter, Narrative Analysis, November 12, 1965.

subsystem to be integrated with the flight spacecraft only at Cape Kennedy facilities, very late in the prelaunch schedule of activities.[59]

The Status of the Boeing Contract

While Boeing and NASA Lunar Orbiter management took steps to improve the delivery schedules at the subcontractor level, Scherer's office was becoming more anxious about the total effect which the various hardware, management, and funding problems could have upon the incentive provisions of the Boeing Lunar Orbiter contract. In the original contract, signed May 7, 1964, the target cost for the entire program had been $75,779,911. The target fee had been $4,736,244. The contract stated explicitly that "in no event shall the sum of the fee, adjusted pursuant to paragraphs (b) and (c) below, be more than fifteen percent (15%) of target cost nor less than zero percent (0%) of target cost."[60] Paragraph (b) further stipulated how the actual cost was to be established and how the target fee was to be revised. Explicitly the contract read: "(A) If the cost is equal to the target cost, the fee to be paid shall be the target fee.

---

[59] Ibid.

[60] National Aeronautics and Space Administration Negotiated Contract No. NAS 1-3800, May 7, 1964, Part II, Fee Incentives, p. 1.

(B) If the cost is less than the target cost, the fee to be paid shall be increased by ten percent (10%) of the amount by which the cost is less than the target cost. (C) If the cost is greater than the target cost, the fee to be paid shall be decreased by ten percent (10%) of the amount by which the cost is greater than the target cost."[61]

The crucial part of the Lunar Orbiter incentive-fee contract hinged upon the provisions defining the incentives. Two specific items determined the incentives: delivery and performance. An Evaluation Board composed of the Associate Administrator of the Office of Space Science and Applications the Director of the Langley Research Center (or their nearest and equivalents) and a chairman appointed by the Associate Administrator of NASA, would be responsible for evaluating the contractor's performance and delivery of the spacecraft in accordance with predetermined schedules. The contract stated that NASA would penalize the contractor "up to a maximum of $10,000 for each individual delivery date, for each calendar day, including Saturdays, Sundays, and holidays, by which actual accomplishment of delivery and acceptance shall have been later than the target date as set forth below. Spacecraft deliveries to the National Aeronautics and Space Administration will be effected in a sequential manner as follows:

---

[61] Ibid., p. 2.

Flight Spacecraft No.	Delivery Date
1	May 7, 1966
2	May 7, 1966
3	July 21, 1966
4	October 21, 1966
5	December 18, 1966"[62]

These provisions were tempered by two other stipulations that held the reduction in fee for any individual delivery to a maximum of $300,000, the equivalent of a delivery thirty days late. Moreover, the total penalty for all delays or late deliveries resulting from "causes beyond the control and without the fault or negligence of the Contractor as defined in Clause 12, Excusable Delays (September 1962), of the General Provisions attached hereto," was the responsibility of NASA.[63]

The history of the Lunar Orbiter Program until the last quarter of 1965 showed several constraints which possibly threatened delivery and over which Boeing had little or no control. The funding situation has previously been discussed as one of these constraints. Another one was the failure of NASA to couple delivery of ground spacecraft with flight spacecraft in the incentive provision of the contract. This failure created an awkward situation by October, which Scherer outlined in a memorandum to Clifford H. Nelson and Sherwood L. Butler at Langley. As certain hardware difficulties, the V/H sensor and the 610-mm-focal-length camera

---

[62] Ibid.
[63] Ibid., p. 3.

lens shutter for example, caused delays stretching into weeks, the testing programs for the ground spacecraft suffered. However, these delays did not hold up fabrication, testing, and delivery of flight spacecraft because, as defined by the contract, the flight spacecraft could be delivered to NASA without the contractor having performed adequate prototype testing.

Thus, the delivery schedule incentive was in danger of losing its meaning. In fact, this condition in the contract's structure--allowing flight spacecraft deliveries without their being contingent on the development and testing of ground spacecraft--constituted a major loophole for Boeing, and Scherer urged that Langley Research Center compensate for it immediately.[64]

Scherer pointed out that when the time came for the three-man Evaluation Board to perform its tasks, the contractor would naturally be prepared to offer "the strongest possible justification of schedule delays based on government actions, such as late government furnished equipment or facilities and conflicts that will likely develop between Orbiter and other programs in the DSN."[65] It was absolutely necessary for the Lunar Orbiter Program to substantiate the

---

[64] Memorandum from Manager, Lunar Orbiter Program, to Langley Research Center, Attention Mr. C. H. Nelson and Mr. S. L. Butler, October 28, 1965.
[65] Ibid., p. 1.

arguments of the Evaluation Board with verified documentary evidence pertaining to all aspects of the incentive provisions in the contract.

Spacecraft Compatibility with Launch and Tracking Facilities

On April 20, 1965, representatives from Boeing, Lockheed, Langley, JPL, and Goddard Launch Operations had met at Kennedy Space Center for a major status review of the spacecraft and the preliminary mission plans. Boeing had presented its plans for using the Eastern Test Range facilities to conduct compatibility tests with a ground spacecraft. At this time it had also requested that it be allowed to evaluate checkout and operating procedures at ETR with the spacecraft's compliance to range requirements. This request necessitated the use of a launch vehicle, which the Lewis Research Center was to supply through Lockheed.[66] NASA approved Boeing's request.

As part of the evaluation, Boeing and Lockheed coordinated their efforts with the Goddard Launch Operations facility, Greenbelt, Maryland, to develop spacecraft flow data for Launch Complex 13 at Cape Kennedy. They completed this activity by May 10. NASA and Boeing further evaluated the requirements of the Deep Space Instrumentation Facility and

---

[66] Boeing Quarterly Technical Progress Report, April to June 1965, Section IV, p. 64.

the Space Flight Operations Facility, whose stations around the world would be used in Lunar Orbiter flight operations. On June 16 Boeing and Eastman Kodak officials met with personnel of the DSN to establish the interface between Eastman Kodak equipment and the DSN. Once this was completed Boeing assisted the DSIF in the development of an activation plan for flight operations. The Deep Space Network was to concur on the plan before it could be implemented.[67]

During the remainder of 1965 and the first half of 1966 major reviews took place in all areas of the Lunar Orbiter Program: spacecraft subsystems, testing and integration with launch facilities, and compatibility with Apollo and Surveyor requirements. The Deep Space Network, meanwhile, had committed the Goldstone Echo site (DSIF 12) to the Lunar Orbiter Performance Demonstration Test throughout 1965. During this time Spacecraft C was given basic compatibility tests to check its systems design with the DSN.[68]

One thorny problem was left to threaten the completion of Lunar Orbiter testing at Goldstone. The Pioneer Mission A had placed a claim on Goldstone facilities that

---

[67] Ibid., pp. 65-66.

[68] Memorandum from Lunar Orbiter Program Engineer, Leon Kosofsky, to Lunar Orbiter Operations Working Group (SL), Subject: Potential Conflict in Goldstone Support of Lunar Orbiter Performance Demonstration Test and Pioneer Mission A, November 22, 1965.

required that the DSN station provide "coverage of one pass per day for each of the first 30 days after launch."[69] Moreover, Goldstone would track the Pioneer space probe on one pass per day for three days a week for the time of launch plus thirty days to six months--a substantial amount of time, impinging on the Lunar Orbiter Performance Demonstration Test still in progress.

The period from December 13, 1965, to February 3, 1966, had been designated by Boeing for the final test phase. Once Spacecraft C had finished the Goldstone tests, it would be shipped to Cape Kennedy for further tests in the Hangar S facility. As things stood the Pioneer launch threatened to delay Spacecraft C in the Goldstone tests, and this was something over which Boeing had no control. Thus a delay here would be charged to NASA's account in the final evaluation of whether the contractor met the incentive requirements of the contract.

Kosofsky made the Flight Operations Working Group aware of the potential conflict and requested that it strive to minimize any delays in the Performance Demonstration Test. Some testing of the Lunar Orbiter could be conducted at Hangar S with Spacecraft 3, but it would lack the photographic subsystem.

---

[69] Ibid.

The situation at the Deep Space Network was the result of scheduling within the Office of Space Science and Applications, which held the responsibility for Lunar Orbiter, Surveyor, Mariner, and Pioneer and their use of the DSN facilities. The DSN did not overcommit its available time or facilities; instead it had to play the juggler, compensating for the schedule slippages in the various programs which relied on DSN. Marshall Johnson, DSN Manager for Lunar Orbiter, attempted successfully to rectify the time-sharing, computer-sharing needs of each program and thus avoided an impact on Lunar Orbiter's schedules.[70]

While Johnson took action at the DSN with the Surveyor, Mariner, and Pioneer projects to compensate for real and anticipated schedule slippages, Scherer continued to prod Eastman Kodak and its subcontractor Bolsey to meet their schedule delivery dates. In a brief memorandum to Oran W. Nicks he explained that he, Clifford H. Nelson, and Eugene Draley at Langley had conferred on the status of the EK/Bolsey situation. They had recommended to Floyd L. Thompson, Langley Director, that Thompson talk to Eastman Kodak management officials by telephone about the schedule situation instead of paying them a top-level visit.[71]

---

[70] Letter, Rechtin to Emme, November 18, 1969.

[71] Memorandum from SL/Manager, Lunar Orbiter Program, to SL/Director, Lunar and Planetary Programs, March 7, 1966.

In addition to Scherer's recommendation, Newell, NASA Associate Administrator for Space Science and Applications, notified NASA Deputy Administrator Seamans early in March of the Lunar Orbiter Program's schedule difficulties.

Newell asked Dr. Seamans to release a telegram to the Boeing Company in an effort to bring the continual series of small schedule slips under control before they escalated into a costly launch delay. The telegram, released by Seamans on March 10, was addressed to Vice President Lysle Wood at Boeing. Showing top-level concern at NASA Headquarters over the threatened status of the Lunar Orbiter schedules, it read:

> The schedule of lunar orbiter is one of the highest priority to NASA. Both unmanned and manned lunar landing missions need the data to be obtained from successful lunar orbiter missions in order that our lunar exploration program can proceed as planned. Scheduled launch dates are requiring firm commitments for world wide network operations. Severe conflicts and delays may occur unless these launch dates can be adhered to.
>
> In view of these facts I have become very concerned about the pattern of delays in deliveries of certain items for the orbiter, such as the photographic system and the inertial reference unit.
>
> I want to emphasize the national importance of this program, the necessity for firm schedule adherence, and to inform you of my personal interest and concern in this matter.[72]

---

[72] Memorandum from S/Associate Administrator for Space Science and Applications to AD/Deputy Administrator, March 9, 1966, with telegram attached.

Seamans indicated in his telegram to Boeing the kind of collision between various programs dependent upon the same facilities which delays could cause. Early in April 1966 further minor delays in deliveries of the photographic subsystem occurred. There had been film alignment problems on the first flight-configured photo subsystem, delaying delivery by one week. The V/H sensor in the first flight-unit photo subsystem had developed troubles which threatened to delay the delivery of this vital component until June 15. To compensate for this Boeing recommended that the V/H sensor from Spacecraft 2 be substituted on Spacecraft 4. This change would ensure delivery of the first flight spacecraft by June 1, but it would reduce the time for the mission simulation testing of the photo subsystem on Spacecraft 2. Yet under the existing constraint of a July launch it was the best alternative.[73]

Flight Spacecraft 4, the first Orbiter destined for the Moon, was undergoing match-mate with the adapter and the shroud at Boeing by April 7. Boeing would subject it to vibration and thermal vacuum tests which it would complete on April 19. Then, if all went well, Boeing would ship it to NASA facilities at Cape Kennedy by May 10. Complementing these tests were two other items that had reached successful completion: the software demonstration tests (i.e., computer

---

[73] Memorandum from SL/L. R. Scherer, to SL/O. W. Nicks concerning update of Orbiter status, April 7, 1966.

219

programming for flight trajectory analysis and tracking) and inter-station compatibility tests. These activities led to the next major item on the schedule: formal mission simulation tests, which were due to begin on April 11.[74]

Flight Recording Equipment

On April 4 Leonard Reiffel of the Apollo Program notified Oran W. Nicks that Apollo requirements for Lunar Orbiter data made it highly desirable, if not necessary, to have sufficient magnetic recording facilities to record incoming data on magnetic tape. He stated that quantitative photometric work made the use of magnetic tape superior to film because: "1. the quality of the data is degraded in the ground photographic process, and 2. magnetic tape provides higher data processing convenience and speed."[75]

---

[74] Ibid.

[75] Memorandum from MA-6/L. Reiffel to SL/O. W. Nicks, Subject: Project Apollo Requirements for Lunar Orbiter Data, April 4, 1966. See also Bellcomm Technical Memorandum 65-1012-6, "Tape Recording of Lunar Orbiter Pictures," by C.J. Byrne, July 6, 1965. Recording on film of raw data transmitted by Lunar Orbiter presented certain limitations. First, film had a very limited dynamic range and did not lend itself easily to enhancement. Second, it was much more difficult to computerize data from a film source than from magnetic tapes. Data recorded on tapes were the direct input signals from the spacecraft. This method of recording also eliminated any film processing errors and provided a greater dynamic range for analytical purposes. Once the tape-recorded data were computerized they could be enhanced by eliminating known and suspected interferences before reconstructing the pictures of the lunar surface with such detail that slopes could be accurately determined within the constraints of Apollo requirements. Film-recorded data did not afford this flexibility

Reiffel emphasized the necessity to have back-up recorders to record all data and avoid irretrievable losses. If, however, this were not possible, he suggested that a tape change schedule be set up which would allow tapes on primary recorders to be changed during times when low-resolution frames were being received at Deep Space Network facilities. He further requested of Nicks a firm commitment on the availability of recorders, including those for the first mission. He stressed that Apollo site selection analysis depended heavily on magnetically recorded data, and he requested more specific information on the Lunar Orbiter Program's plans for automatic data processing and validity tests of processed data.[76]

Nicks replied to Reiffel's memorandum on April 26. He concurred that a meeting between technical specialists from both programs should be called to discuss the problem of magnetic recording of data, the availability and cost of extra recorders, and the best way to secure Lunar Orbiter data in a form that the Apollo Program could use at the earliest possible date. He also pointed out that the Deep Space Network had received three Ampex FR 900 recorders but that their necessary amplifiers would not be delivered before June 1. This late delivery, the period of installation and testing, and the training of personnel to operate the

---

[76]*Ibid.*

recorders kept the Lunar Orbiter Program from making a firm commitment to Reiffel for the first flight.[77]

Nicks stated that the problem of back-up recorders had been investigated and the results showed that the contractor, Ampex, could deliver three units by the end of October if an order were placed by May 15, 1966. The earliest date for their operation would be February 7, 1967, and the estimated cost would be about $600,000. Until the Lunar Orbiter Program had more reliable information on the performance of the FR 900 in the field, Nicks did not believe it was advisable to ask the Deep Space Network to purchase additional recorders. However, Boeing had been investigating the feasibility of changing tapes during reception of low-resolution data, and it had indicated that this probably could be done.[78]

A Change in Delivery Incentive

Other areas of major concern existed. One was in the NASA-Boeing contract and the funding relationship. During March and April 1966, the Lunar Orbiter Project Office at Langley negotiated a new delivery incentive with the Boeing

---

[77] Memorandum from SL/Director, Lunar and Planetary Programs, to MA-6/L. Reiffel, Apollo Program Office April 26, 1966.
[78] Ibid.

Company because of the necessity of moving the first launch date from early June to mid-July. The new delivery date was June 20, and the change relieved some of the pressure that schedule delays, especially on the photographic subsystem, had caused in the timetable. In addition NASA officials had taken the opportunity to correct previous weaknesses in the incentive clause of the contract.[79]

Scherer reported to Nicks on April 7 that the Lunar Orbiter Program was close to meeting its obligations according to plan, but that accrued costs were about $10 million behind the plan. The completion costs for RCA were expected to end up one half to one million dollars below the level planned. In addition the Machinists' Union at Boeing had not reached a new contract settlement with the company by the April 7 deadline, and a strike appeared likely. If the union struck before April 30, negotiations would move to Washington, D.C. A strike would affect Lunar Orbiter operations at Cape Kennedy.[80]

Langley had reported to Headquarters at the end of March that the program was proceeding toward a launch readiness

---

[79] Memorandum, Scherer to Nicks, April 7, 1966.
[80] Ibid.

date of July 11, 1966, despite several technical problems
that continued to hold up testing. The major problems
were in the photographic subsystem. The shutter mechanism
for the 610 mm lens and the V/H sensor had not yet been
perfected, and their absence was delaying vital tests of
the subsystem at the flight spacecraft level.[81] The problem continued to persist almost to the actual launch date.
Indeed, the July launch date had to be canceled because the
photographic subsystem was not available, and it was not
until the second week in August that the program was able to
launch a spacecraft.[82]

---

[81] *Project Lunar Orbiter, Narrative Analysis*, Langley Research Center, April 22, 1966.

[82] Taback interview, July 7, 1970.

CHAPTER IX

MISSIONS I, II, III: APOLLO SITE SEARCH AND VERIFICATION

Preparations for the First Launch

NASA launched five Lunar Orbiter spacecraft to the Moon between August 1966 and August 1967, and all five successfully performed their missions. This record set a precedent in the Office of Space Science and Applications in lunar exploration. Not every Orbiter proved an unqualified success, but each one obtained valuable photographic data that subsequently aided the Apollo Program in site selection for the manned lunar landings of Astronauts Neil A. Armstrong and Edwin E. Aldrin, Jr. (Apollo 11, July 20, 1969); Charles Conrad, Jr., and Alan L. Bean (Apollo 12, November 19, 1969); and later missions. Moreover, Lunar Orbiter photos enabled Surveyor Program personnel to verify landing sites and to place Surveyors in highly significant areas on the Moon's surface to perform their missions.

One major reason for the impressive record of five successful missions was the philosophy motivating the many individuals in the program. The men who had spent long months of preparation and training for the Lunar Orbiter flights had developed emergency procedures for many nonstandard situations which might arise. It was, however, obviously impossible to anticipate or simulate all possible failure modes in these training exercises, and only a limited set of contingencies were practiced. The experience gained from these sessions proved invaluable in detecting and eliminating "bugs" in the operational systems, improving detection and correction of potential catastrophes during a mission and the probability of squelching problems in their embryonic stages.[1]

NASA and Boeing had designed Lunar Orbiter to be "tweaked." It was not launched and sent on its way to the Moon and then left alone to perform its mission automatically and expire. On the contrary, it was designed to operate with the assistance of ground controllers to overcome risks in each mission, potential failures in subsystems, and the external hazards of space. Built to function for a thirty-day minimum lifetime and an extended period of operation after the termination of the photographic mission, each of the five Lunar Orbiters proved successful in fulfilling its mission assignments.

The missions, in addition, proved the usefulness of the orbiter concept in unmanned lunar and planetary exploration. Lunar Orbiter--unlike Ranger, which was designed to send back television pictures of the Moon as it raced toward a terminal impact point on its surface--had the greater

---

[1]Memorandum, Crabill to Emme, December 9, 1969, p. 2.

advantage of orbiting its target for an extended period. Ground control operators thus had time to analyze any problems which arose and to prepare commands to the spacecraft to solve each problem.[2] Although risk was a constant companion, the Lunar Orbiters had a new dimension of flexibility once they were in orbit around the Moon. The greatly extended time of an orbiting mission over an impact mission allowed flight operations personnel the luxury of compensation if a command was wrong or sent at the wrong time.

Twenty-eight months of industrious work and planning since the time when NASA Administrator James E. Webb had officially approved the program brought all activities to the eve of the first launch. During the months from April to August 1966 Langley and Boeing completed the final tasks which preceded the launch. On July 25 program officials conducted the Flight Readiness Review at Cape Kennedy, and on July 26 Langley accepted the spacecraft from Boeing, certified ready for launch.[3]

---

[2] Interview with Lee R. Scherer, Program Manager, at Cape Kennedy, July 31, 1967. This was part of a discussion between various members of the Lunar Orbiter Program--including Clifford H. Nelson, Israel Taback, A. Thomas Young, Robert P. Bryson, Dr. Martin Molloy, and the author--at the home of Mrs. Mary Bub, a journalist, in Cocoa Beach, Florida.

[3] Project Lunar Orbiter, Narrative Analysis, Langley Research Center, August 3, 1966.

The First Launch

The launch of *Surveyor I* on May 31, 1966, and its need of the Deep Space Network, together with delivery problems of the photographic subsystem for the first flight Lunar Orbiter at Eastman Kodak, caused the tentative July 11 launch date to be slipped to August 9. By August 1 the photo subsystem had arrived and had been installed on board *Lunar Orbiter I*. On August 2 the spacecraft was transferred to Launch Pad 13 and mated with the Atlas-Agena launch vehicle. Following the mating, project personnel tested the compatibility of the spacecraft with the DSIF Station 71 at the Cape.[4]

On August 9 the Boeing-Lockheed-NASA team at the Eastern Test Range Launch Complex 13 and at support facilities near Hangar S counted the spacecraft down to T minus seven minutes. Then, with the launch only a short time away, an anomaly in the Atlas Propellent Utilization System caused a postponement of the mission until the launch window of the following day.[5]

Lunar Orbiter I, weighing 853 pounds, roared into space atop the Atlas-Agena D launch vehicle at 19:26 Greenwich Mean Time on August 10. Launch operations personnel injected the Agena and the spacecraft into a parking orbit

---

[4] Ibid.

[5] Boeing Quarterly Technical Progress Report, Lunar Orbiter Program, July to September 1966, Section IV, p. 35.

at 19:31 GMT, and at 20:04 the Agena fired its rocket once more to inject the Lunar Orbiter into a trajectory toward the Moon.[6] Lunar Orbiter I deployed its solar panels and antennas as planned and acquired the Sun (the first celestial reference for establishing cruise attitude). The mission continued exactly according to the preflight plan until the time of initial acquisition of the second celestial reference, the star Canopus.[7]

The Canopus star tracker sensor proved to be one of two major problems during the Earth-Moon transit of the spacecraft. On August 11 at 02:14:57 GMT, flight operations personnel at the Deep Space Network facilities at JPL commanded the Canopus sensor to turn on. When it did, it indicated excess voltage, 1.5 times stronger than the preflight calculated signal voltage. Acquisition of Canopus failed. The reason for the failure was thought to be excess light reflected from some part of the spacecraft's structure, stimulating undue response from the sensitive sensor. This problem should have been detected during system testing, but it had not been. However, flight operations attempted a number

---

[6] Ibid., p. 36.

[7] The Boeing Company, Lunar Orbiter Final Mission Report, Vol. III, Mission Operational Performance, Boeing Document D2-100727-3, p. 6.

of tests and experiments to correct or circumvent the anomaly.

The necessity for an attitude-stabilized spacecraft like Lunar Orbiter to acquire proper stabilization in reference to the Sun and the star Canopus cannot be overstressed. Unlike a spin-stabilized spacecraft, Lunar Orbiter I depended on proper orientation along its yaw, pitch, and roll axes to arrive in the Moon's vicinity in the correct attitude to be injected into lunar orbit. After the failure of the Canopus sensor to acquire a fix on Canopus, flight operators were able to save Lunar Orbiter I's mission by developing an alternate procedure. At the time of the midcourse maneuver, they commanded the spacecraft to establish a roll reference by pointing the Canopus sensor at the Moon.[8]

This maneuver was executed successfully and, after the sensor locked on the Moon, the flight controllers were reasonably sure that it was operating correctly. They developed a procedure that used the Canopus sensor during periods of occultation of the Sun to verify or correct the spacecraft's orientation.[9]

The other major problem encountered during the cislunar journey was overheating of the spacecraft. This did not

---

[8] Kosofsky interview.
[9] Boeing, Lunar Orbiter I Final Mission Report, III, p. 6.

become serious until after the midcourse maneuver. To perform this manuever despite the trouble with the Canopus star tracker, Lunar Orbiter flight operators used the Moon as the roll reference. The midcourse maneuver was executed to correct the spacecraft's translunar trajectory in preparation for deboosting it into orbit around the Moon. A second manuever was executed to orient the spacecraft 36° off-Sun for a period of 8.5 hours.[10] The purpose of this move was to lower the spacecraft's temperature on the equipment-mounting deck during transit.

The coating on the exterior of the deck was degrading under solar radiation at the expected rate, and no acute overheating was experienced until Lunar Orbiter I was already in orbit around the Moon. Nevertheless, the planned heat dissipation period when the spacecraft was flown 36° off-Sun did not seem to retard overall degradation of the thermal coating on the exterior of the equipment deck.

The need to regulate the spacecraft's temperature and to investigate the Canopus sensor anomaly necessitated pitch and yaw manuevers every few hours. These added small accelerations to the spacecraft, all approximately in the same direction. Their effect on the prediction of the spacecraft's position at the time of deboost was minimal, and the flight operators successfully worked around the effects of the per-

---

[10] Ibid., p. 7.

turbations resulting from the off-Sun maneuvers. The position of Lunar Orbiter I at the time of the deboost maneuver into initial orbit around the Moon was estimated to b less than ten kilometers off the planned insertion point and presented little difficulty for flight controllers.[11]

Controllers began a series of commands at 15:22:56 GMT on August 14 to place the spacecraft in orbit. Before insertion the spacecraft executed another thermal relief maneuver, which lasted 7.5 hours. The maneuver provided the optimum temperature conditions before the critical insertion. The final sequence of commands for insertion was carried out without any problems, and Lunar Orbiter I was ready to begin the major work of its mission.[12]

The photographic mission of Lunar Orbiter I was entirely Apollo-oriented.[13] Once the spacecraft had been placed in its initial orbit, with an apolune of 1,866.8 kilometers and a perilune of 189.1 kilometers, ground control checked out the subsystems. The necessity to fly off-Sun and the extra number of maneuvers required because of the Canopus sensor problem had affected the interrelationships of the spacecraft

---

[11] J. R. Hall, ed., TDS Final Report, Vol. II, Mission A Summary, No. 608-17, Tracking and Data System Report Series for the Lunar Orbiter Project, November 15, 1969, Jet Propulsion Laboratory, Pasadena, California, p. 4-15.

[12] Boeing, Lunar Orbiter I Final Mission Report, III, p. 8. See also Boeing Quarterly Technical Progress Report, July to September 1966, Section IV, p. 36.

[13] Interview with G. Calvin Broome, Langley Research Center, July 19, 1967.

subsystems, and flight controllers had to make compensations, especially in the power system to avoid overtaxing the batteries.

On August 15, during the sixth orbit, ground control successfully commanded Lunar Orbiter I to read out the Goldstone test film. This film, being the leader on the supply of film for the mission, had been pre-exposed and checked out through tests of the readout subsystem at the DSIF station in Goldstone, California, before the mission. The same data were now read out again and compared to the known results of the Goldstone tests in order to check the performance of the readout and communications subsystems on board the spacecraft.

At the time of the Goldstone test film readout the thermal problem became acute. The coating on the exterior of the equipment deck was supposed to radiate excess heat during periods of solar occultation. It did this approximately as predicted, but heat levels continued to rise, probably because of more rapid degradation in the pigment of the coating than had been expected. However, on August 18, during the twentieth orbit, a power transistor in the shunt regulator array failed, with a compensating effect on battery temperatures. The failure placed an extra load of 1.2 to 1.5 amperes on the power system, increasing the battery discharge rate during occultation of the Sun. The extra load meant that the off-Sun angle of 36° could be reduced slightly at the time when sufficient power for readout was required of the power system.[14] The analysis and compensatory action for this problem reflected outstanding flight operations.

After orbiting the Moon for four days and twenty-three hours Lunar Orbiter I began the first operation of its photo subsystem since the readout of the Goldstone test film. Eleven frames were advanced and processed during the twenty-fifth orbit at 12:12:13 GMT on August 18, bringing the unexposed film into position for the first photographic sequence, which was to begin on orbit 26.

The photographic subsystem, which Eastman Kodak had designed and built, was put together with the precision of a Swiss watch. Every component of the subsystem was tightly housed in an aluminum "bath tub" a little larger than a large round watermelon. A precision instrument with a very complex task to perform, the photo subsystem operated like a thrashing machine. The film, which had to go through three plane changes, was drawn from the supply spool, clamped in a movable platten, moved and exposed simultaneously, and advanced farther to make room for a new film--all in a matter of a few seconds.[15]

---

[14] Boeing, Lunar Orbiter I Final Mission Report, III, p. 9.
[15] Costello interview.

The first site to be photographed, Site I-0 (a portion of Mare Smythii), was covered by the Orbiter's dual lens camera as planned. Photo subsystem telemetry to Earth appeared to be normal. The photos were taken as follows. Ground control commanded the spacecraft to open the camera thermal door. Two photo sequences were then executed: one of sixteen frames in the high-resolution mode and one of four frames in the medium-resolution mode. They were made at an altitude of 246 kilometers above the Moon while the spacecraft's velocity relative to the lunar surface was 6,400 kilometers an hour. Exposure time for each shutter was 1/50 of a second, and simultaneous medium- and high-resolution pictures were made every ten seconds. After the sequences, the thermal door was closed and the film was processed.[16]

Five hours later the readout process began, at 19:50:52 GMT on August 18. All the medium-resolution frames were of excellent quality, but reconstruction of four high-resolution frames revealed severe image smearing. The first high-resolution frame contained some unsmeared data, but George Hage, the Boeing Lunar Orbiter Program Engineer, recognized it to be a double exposure. The first exposure

---

[16] Lunar Orbiter Program Office, NASA, Lunar Orbiter I Mission Status Report 8, Status as of 11:30 EST, August 18. 1966. Note: all times for the five missions are given exactly as they appear in the mission status reports. The time used was local time at the site where the mission was being monitored, with the exception of Mission I.

of the frame contained unsmeared data and proved to have been taken prematurely of a feature east of the planned target area when the V/H sensor was turned on.[17] Apparently the shutter of the 610 mm lens was out of synchronization with the V/H sensor; further investigation demonstrated that this supposition was true.[18]

Flight operators in charge of mission photography set up an experiment to examine the possible causes of the smearing. After completion of the Site I-0 photography ten more exposures were made with the 610 mm lens for purposes of evaluating exposure 26, the first picture of the four-frame sequence after photographing Site I-0. One test involved the use of different exposure rates with and without the V/H sensor turned on. A second test was used to determine if, in fact, the V/H sensor was causing abnormal shutter operations. It consisted of three steps:

1) The camera thermal door was opened and the V/H sensor was turned on.

2) The sensor was left on for approximately 2 minutes and then turned off.

3) The camera thermal door was then closed and the camera shutter was commanded to take a picture with the door closed and to move fresh film into the camera for the next photograph.[19]

---

[17] Memorandum from Dennis B. James, Bellcomm, Inc., to Dr. Eugene M. Emme, Subject: Comments on manuscript "Lunar Orbiter: A Preliminary History," November 17, 1969, p. 3.

[18] Lunar Orbiter I Mission Status Report 9, Status as of 9 a.m. EDT, August 19, 1966.

[19] Lunar Orbiter I Photographic Mission Summary, NASA CR-782, April 1967, p. 46.

The second test confirmed that the abnormal operation occurred when the V/H sensor was on; a high-resolution exposure was made with the thermal door open and no shutter command, but no medium-resolution picture was taken when the shutter command was given. Despite the problem, flight controllers made no deviations from the flight plan, and the spacecraft was transferred to its lower, final orbit at 09:49:58 GMT on August 21.[20] The new orbital parameters were: apolune, 1,855 kilometers; perilune, 58 kilometers; inclination to the lunar equator, 12.32°.[21]

Just before the orbit transfer, Lunar Orbiter I took two frames of medium- and high-resolution pictures of the Moon's far side at an altitude of 1,497 kilometers. The V/H sensor was off, because there was no need for image-motion compensation at such a high altitude. After the frames were read out, they revealed high-quality pictures of the lunar surface in both medium- and high-resolution modes, without smearing.[22]

Another problem occurred before the final orbit transfer, requiring the photo subsystem to take additional unplanned photographs. The Bimat apparently was sticking.

---

[20] Lunar Orbiter I Final Mission Report, III, p. 10.

[21] Lunar Orbiter I Mission Status Report 11, Status as of 8:30 a.m. EDT, August 22, 1966.

[22] Ibid.

The original plan had called for fresh Bimat to be placed on the processing drum at least every 15 hours. This meant that two frames would be processed every four orbits. However, evidence of Bimat stick in the early frames precipitated the decision to use additional film which would permit processing during every orbit. Eight extra pictures were to be taken.[23] This change and the extra diagnostic pictures taken to evaluate the high-resolution shutter problem forced a revision in the planned photographic coverage of the remaining sites. The result was that only eight exposures would be taken of Sites 4, 6, and 8, while the other sites would receive the original 16-frame coverage.[24]

The trouble in the high-resolution camera lens shutter continued to plague photography when the V/H sensor was operating, despite the increase in output voltage which Eastman Kodak technicians had recommended during analysis of the problem. Further analysis revealed that the logic-control circuitry of the 610-mm-lens focal-plane shutter was susceptible to electromagnetic interferences which caused it to trip at the wrong part of the image-motion compensation cycle. It was not possible to solve this problem by modifying procedures, and low-altitude high-resolution

---

[23] Lunar Orbiter I Photographic Mission Summary, NASA CR-782, p. 46.
[24] Ibid.

photography on the first mission proved a failure despite further attempts to correct the problem.

Nitrogen gas, used by the attitude control subsystem to manuever the spacecraft, had been expended in greater amounts than originally planned because of the difficulties in the Canopus star tracker and alterations of planned photography caused by the shutter problems and the evidence of Bimat sticking. Moreover, thermal relief maneuvers and excess attitude update maneuvers, together with the failure of a gas regulator, increased the rate of nitrogen usage. Between August 23 and 31 an average of 0.17 kilograms of nitrogen was expended per day. Flight controllers tried an economizing procedure. They commanded the spacecraft to fly off-Sun on its pitch axis and to update its attitude on the pitch and yaw axes using the coarse Sun sensors and on its roll axis using the Canopus sensor. This change resulted in an expenditure of 0.04 kilograms per day between September 1 and 14.[25]

From the final orbit perilune of 58 kilometers, Lunar Orbiter I was deboosted successfully to a lower altitude of 40.5 kilometers for further photography on August 25. This move was the result of an analysis of the V/H sensor in a duplicate Lunar Orbiter photo subsystem on the ground

---

[25] Boeing, Lunar Orbiter I Final Mission Report, III, p. 11.

by Eastman Kodak engineers on August 24. They had concluded that there was a possibility that the camera would operate normally below an altitude of 51 kilometers.[26] They reasoned that, since the ratio of velocity to height would be higher in the new, lower orbit, the image-motion compensation mechanism might be forced into synchronization with the 610 mm lens's focal-plane shutter. Synchronization was, unfortunately, never attained, but there was some reduction in smearing because a higher solar lighting angle permitted a change in shutter speed from 1/50 to 1/100 second.[27]

By August 29 <u>Lunar Orbiter I</u> had completed its photographic acquisition, with a total of 205 exposed frames. Of these, 38 frames had been taken in the initial orbit; 167 were made in the lower orbits. The spacecraft photographed all nine potential Apollo landing sites. Pictures of eleven sites on the far side of the Moon and two Earth-Moon pictures were also taken. The complete readout of the photographs began on August 30.[28]

Despite the malfunctions in the photographic subsystem the spacecraft succeeded in taking many historic pictures. Command and maneuver requirements were developed to take,

---

[26] Lunar Orbiter I Mission Status Report 14, Status as of 9 a.m. EDT, August 24, 1966.

[27] Lunar Orbiter I Mission Status Report 18, Status as of 10 a.m. EDT, August 29, 1966.

[28] Lunar Orbiter I Mission Status Report 20, Status as of 11 a.m. EDT, September 1, 1966.

in near real-time, such pictures as those of the morning and evening terminator on the lunar surface, the Earth as seen from the Moon's vicinity, numerous farside pictures, and additional photographs of sites of interest on the near side. _Lunar Orbiter I_ photographed such areas as potential targets for Mission B, major craters, and mare and upland areas useful as Apollo navigation landmarks and was mostly able to satisfy the requirements to take these photographs.[29]

    Of all the pictures which _Lunar Orbiter I_ made, one of the most spectacular was the first photograph of the Earth taken from the vicinity of the Moon. This picture was not included in the original mission plan, and it required that the spacecraft's attitude in relation to the lunar surface be changed so that the camera's lenses were pointing away from the Moon. Such maneuvering meant a calculated risk and, coming early in the flight, the unplanned photograph of Earth raised some doubts among Boeing management about the safety of the spacecraft.

    Robert J. Helberg, Boeing's Program Manager for Lunar Orbiter, opposed such a hazardous, unnecessary risk. The spacecraft would be pointed away from the Moon so that

---

[29] _Lunar Orbiter I Photographic Mission Summary_, NASA CR-782, p. 46.

the camera's lenses could catch a quick view of Earth tangential to the lunar surface. Then, once the pictures were made (flight controllers would execute two photo sequences on two different orbits), <u>Lunar Orbiter I</u> would disappear behind the Moon where it would not be in communication with ground control. If, for some reason ground control failed to reestablish communications with it, the Apollo-oriented mission photography would probably remain undone. Moreover, Boeing had an incentive riding on the performance of the spacecraft, and Helberg did not think it prudent to commit the spacecraft to a series of maneuvers for which no plans had been made.[30]

The understandably conservative Boeing stance was changed through a series of meetings between top NASA program officials, including Dr. Floyd L. Thompson, Clifford H. Nelson, and Lee R. Scherer. They convinced Helberg that the picture was worth the risk and that NASA would make compensation in the event of an unexpected mishap with the spacecraft. After agreement had been reached, Lunar Orbiter flight controllers executed the necessary maneuvers to point the spacecraft's camera away from the lunar surface, and on two different orbits (16 and 26) it recorded two unprecedented, very useful photographs.

---

[30] Taback interview. See also Transcript of Proceedings--Discussion between Nicks, <u>et al.</u>, and members of National Academy of Public Administration, pp. 111-112.

The Earth-Moon pictures proved valuable for their oblique perspective of the lunar surface. Until these two photographs, all pictures had been taken along axes perpendicular or nearly perpendicular to the Moon's surface. On subsequent Lunar Orbiter missions oblique photography was planned and used more often.[31]

<u>Lunar Orbiter I</u> began its extended mission on September 16 after completion of photographic readout. During this period non-photographic data was telemetered to Earth at regular, planned intervals. Flight controllers monitored the orbital behavior of the spacecraft, the micrometeoroid detectors, and the condition of the power, attitude control, and communications subsystems.

By October 28 the condition of <u>Lunar Orbiter I</u> had deteriorated significantly. Scherer issued a status report which pointed out the following: 1) very little gas remained for attitude control (0.4 kilograms at 7 kilograms per square centimeter--100 psi--pressure); 2) estimated stabilized life of spacecraft was two to five weeks; 3) the battery was losing power because of prolonged overheating, and if it fell below 15 volts, the onboard flight programmer would lose essential

---

[31] For a detailed technical description of the Earth-Moon photographs refer to <u>Lunar Orbiter I--Photography</u>, NASA CR-847, prepared by the Boeing Company, Seattle, Washington, for the Langley Research Center, August 1967, pp.64-71.

parts of its memory; 4) the transponder was responding erractically, and the inertial reference unit was losing its ability to keep the spacecraft stable. The program manager and his staff realized that loss of control over communication transmission from Orbiter I could jeopardize the mission of the second Lunar Orbiter. They conferred with members of the Langley Lunar Orbiter Project Office who, in turn, decided to command the spacecraft to impact on the far side of the Moon during its 577th orbit on October 29. This maneuver, successfully executed, brought the first mission to an end.[32]

Results of the First Mission

Lunar Orbiter I photography was subjected to numerous analyses, photometric enhancement processes, and evaluations by technicians and scientists at the Langley Research Center. Following this a more extensive screening process of Mission I photography was made by specialists from Jet Propulsion Laboratory, the Manned Spacecraft Center, NASA Headquarters, Boeing, the United States Geological Survey, and Langley. They studied very carefully all Orbiter I photographs and generated preliminary terrain and geologic maps and screened photographic data for acceptable Apollo sites.

---

[32] Memorandum from SL/Manager, Lunar Orbiter Program, to the File, October 28, 1966, Subject: Lunar Orbiter I situation. See also Astronautics and Aeronautics, 1966, NASA SP-4007, Washington, D.C., 1967, pp. 332-333.

This effort started the major process of Apollo site selection and data analysis.[33]

Some of the most significant problems which the first mission photography revealed were the following: 1) photographic imperfections due to mechanical operation in the photo subsystem (for example, partial dryout of the Bimat because of pressure variation of a roller in the processor mechanism produced a narrow strip of incorrectly processed film); 2) density variations caused by the Ground Reconstruction Equipment kinescope tubes; 3) smear of high-resolution photographs caused by inadvertent triggering of the focal-plane shutter of the 610 mm lens. This problem has been previously discussed.[34]

Prelude to Mission II

At the time of launch of <u>Lunar Orbiter I</u> the status of the other spacecraft was as follows. Spacecraft 5, the second flight spacecraft, was in storage at Cape Kennedy. Its photo subsystem was due to be delivered at KSC on September 4, 1966. Spacecraft C, a ground-test spacecraft, was at JPL for display purposes, and no further work was planned for it. Spacecraft 1, also a ground-test spacecraft,

---

[33] Langley Working Paper: Preliminary Terrain Evaluation and Apollo landing site analysis based on Lunar Orbiter I Photography.

[34] <u>Lunar Orbiter I--Photography</u>, NASA CR-847, pp. 11-17.

was at Boeing in Seattle. It had completed formal testing and was being used as a flight-test unit. During Mission I Boeing used it to duplicate problems encountered on Lunar Orbiter I as an aid to their resolution. Spacecraft 2 was also at Boeing, awaiting its photography subsystem so that it could begin mission simulation tests. Spacecraft 3, the fifth flight spacecraft, was in the clean room at Boeing waiting for various hardware components to be installed. Major testing of this spacecraft was due to begin on November 7. Spacecraft 6, the third flight spacecraft, was scheduled for preshipment review on August 19 followed by shipment to Cape Kennedy on August 20. Spacecraft 6 would then serve as a back-up for the second flight spacecraft. Finally, Spacecraft 7, the fourth flight spacecraft, was in storage at Boeing awaiting preenvironmental flight checkout, scheduled to begin on August 29.[35]

The second Lunar Orbiter mission had run into difficulties during May 1966, six months before the tentative November launch date for Lunar Orbiter II. On May 20 NASA and Boeing program officials conducted a preshipment review of Spacecraft 5 at the Boeing Company. This spacecraft was to serve as back-up for the first mission and was to be launched on the second mission in the event that all went as planned

---

[35] Project Lunar Orbiter, Narrative Analysis, Langley Research Center, August 17, 1966.

on the first. After reviewing the history of Spacecraft 5, NASA's review team refused permission to ship it to Cape Kennedy facilities without further testing.[36] The Boeing Lunar Orbiter Program officials objected to this, but the history of Spacecraft 5 revealed a need to overcome inadequate operations of important equipment.

Having been subjected to the same tests as Spacecraft 4, Spacecraft 5 was considered ready for shipment with one major exception. The camera thermal door had failed to open during thermal vacuum testing. The other thermal vacuum tests were completed, save for this one. Again it was attempted. The thermal vacuum chamber was pressurized and the command for the door to open was sent. Again it remained closed. Next the operation of the thermal door was visually observed, and after some of the thermal insulation had been pulled loose the door operated correctly through several cycles. The door and its motor mechanisms were then removed from the spacecraft for special thermal vacuum tests.[37]

Boeing officials wanted to ship the spacecraft to Cape Kennedy without the door while it underwent further tests.

---

[36] Memorandum from SL/Manager, Lunar Orbiter Program, to the File, May 24, 1966, Subject: Preshipment Review of Second Lunar Orbiter Flight Spacecraft. (The NASA review team consisted of Lee R. Scherer, Clifford H. Nelson, Israel Taback, Kenneth L. Wadlin, James B. Hall, and Messrs. Jackson and Eckhard.)

[37] Ibid.

Once the cause of failure was isolated, it could be corrected, and the door could be reinstalled at the Cape. NASA officials declined this suggestion because of the long history of development troubles with the door mechanism. Nevertheless, Boeing officials still wanted to ship the spacecraft, saying that they would be merely effecting a transfer from Boeing-Seattle to Boeing-Florida. Boeing's major reason was the delivery deadline for the second flight spacecraft: June 22. A contract incentive depended upon meeting this date. However, NASA officials still disagreed with Boeing's line of reasoning and insisted that the facts were clear. The spacecraft had failed a specified test. It was necessary to retest the whole spacecraft. Reluctantly Boeing management accepted this verdict and issued instructions to return the spacecraft to the test chamber on May 21.[38]

The Plan for Mission II

While Boeing reworked the camera thermal door, the Lunar Orbiter Project Office at Langley continued to formulate plans for the second mission. Original planning for Mission B had only photographic data from Earth-based telescopes and Ranger spacecraft to rely upon because <u>Lunar Orbiter I</u> had not yet flown. On May 6, 1966, representatives

---

[38] Ibid., p. 3.

from Bellcomm and the Apollo, Surveyor, and Lunar Orbiter Program offices convened at Langley for the Mission B Planning Meeting. The information and requests which they provided enabled Langley mission planners to set up the following guidelines for Lunar Orbiter Mission B:

1. Distributed sampling with a string of sites in the northern part of the Apollo zone.

2. Sampling of both mare and highland with greatest number of samples in the mare.

3. Sites spaced consistent with the lighting of LEM landing constraints. (Present value of sun elevation of 7 to 20 degrees would be used, resulting in optimum spacing equaling 11 degrees, plus or minus 2 degrees.)

4. One of the mare sites to be the Ranger VIII impact point.

5. The availability of a landed Surveyor or any new data to necessitate a review of any mission design.

6. Mission B sites to be selected whose terrain to the east appeared to be consistent with the Apollo landing approach constraints, where possible.[39]

The members of the several organizations at the meeting aided Langley officials in producing a Mission B plan which the Lunar Orbiter Program Office in Washington presented to the Surveyor/Orbiter Utilization Committee on June 1. The plan had three primary goals based upon Ranger and Earth-

---

[39] Minutes of the Lunar Orbiter Mission B Planning Meeting, Langley Research Center, May 6, 1966 (recorded by A. Thomas Young), pp.5-6.

telescope data and performance evaluations of the Lunar Orbiter spacecraft subsystems:

    A.   Photographic -- To obtain detailed lunar topographic and geologic information of various lunar areas to assess their suitability for use as Apollo landing sites.

    B.   Selenodetic -- To provide trajectory information which will improve the definition of the lunar gravitational field.

    C.   Environmental -- To provide measurements of micrometeoroid and radiation flux in the lunar environment for spacecraft performance analysis.[40]

Apollo requirements had priority as on the first mission. The area to be covered was a swath along the front side of the Moon ranging from +5° to -5° latitude and +45° to -45° longitude. Topographic considerations affecting the mission plan dictated that Lunar Orbiter B (*Lunar Orbiter II*) look for areas smooth enough for the Apollo Lunar Module to land on. The approaches to these areas had to be free of obstacles over a certain height to allow satisfactory performance of the Lunar Module landing radar.[41] Because the Apollo missions would operate in a retrograde lunar orbit instead of the posigrade orbit of the Lunar Orbiter missions, the landing approach zone would be east of the

---

[40] Lunar Orbiter Project Office, Langley Research Center, Lunar Orbiter Mission B Description, June 1, 1966.

[41] *Ibid.*, p. 7.

landing site.[42]

The Lunar Orbiter Project Office at Langley selected eleven sites pertaining to Apollo missions to be photographed on the second Orbiter mission. In order to keep the mission simple the spacecraft would execute a minimum number of attitude maneuvers. There would be one photographic pass per site, and high orbit photography would be eliminated. Lunar Orbiter II would carry out contiguous high-resolution vertical photographic coverage between adjacent orbits. This called for an inclination of 11° to 12° to the lunar equator. Surface lighting conditions had to be such that photography could detect cones of two-meter diameter and one-half meter height and slopes of 7° in an area of seven meters square.[43]

On September 29 the tentative Mission B plan was amended. The photography and spacecraft performance evaluations of Lunar Orbiter I--in addition to further inputs from Bellcomm, the U.S. Geological Survey, the Army Map Service, the Manned Spacecraft Center (Houston), NASA Headquarters Office of Manned Space Flight, and the Surveyor Project Office--confirmed tentative mission objectives for the second Lunar Orbiter flight more than they altered them. As of October

---

[42] Apollo had to operate in a retrograde orbit--that is, an orbit whose direction was counter to the rotation of the Moon--in order to have the safety option of a free Earth-return trajectory in case of an emergency such as occurred later on Apollo 13 in April 1970. Lunar Orbiter operated in a posigrade orbit--that is, in the direction of the Moon's rotation--because it did not have to plan for this contingency.

[43] Lunar Orbiter Mission B Description, p. 12.

these objectives were:

> <u>Primary</u> -- To obtain, from lunar orbit, detailed photographic information of various lunar areas, to assess their suitability as landing sites for Apollo and Surveyor spacecraft, and to improve our knowledge of the Moon.
>
> <u>Secondary</u> -- To provide precision trajectory information for use in improving the definition of the lunar gravitational field.
> To provide measurements of micrometeoroid flux and radiation dose in the lunar environment, primarily for spacecraft performance analysis.[44]

During the process of site selection for the second Orbiter mission a hypothesis based upon Earth-telescope photography and the very useful <u>Ranger VII</u> pictures exerted a particular influence on the choice of sites. Data from these two earlier sources tended to show that bright rays extending from younger craters were actually heavily cratered, making landings very hazardous or impossible in such areas. To test this, <u>Lunar Orbiter I</u> had photographed sections in lightly rayed areas. Specifically, photographs of Site A-3 in Mare Tranquillitatis revealed smooth areas where a Lunar Module could land. <u>Orbiter I</u> Frame M-100 of Site A-3 showed an area in a light ray where cratering was insufficient to rule it out as a landing site. The ray in this photograph was faint and probably had its origins in

---

[44] Lunar Orbiter Project Office, Langley Research Center, Lunar Orbiter Mission II Description, as amended on September 29, 1966, issued October 26, 1966, p. 3

the crater Theophilus but had subsequently been filled in.[45]

Planners concluded from Orbiter I photography that some ray areas were possibly smooth. Moreover, photography from the first Orbiter had actually previewed certain targets in the second mission. Thus planners decided to change several sites in Mission B and to have Lunar Orbiter II look at the ray areas between the lunar craters Copernicus and Kepler, extending north of the western Apollo zone. The Mission B plan was thus substantially revised as a result of the divergences between Ranger VII and Lunar Orbiter I photographs of crater rays.[46]

The Second Mission

Less than three months elapsed between the launch of the first Orbiter and that of Lunar Orbiter II. On November 6, 1966, the second mission began, with the launch of the spacecraft at 23:21 GMT. The cislunar transit went as planned, with no trouble in the Canopus star tracker. One reason for success was that the solar panels and parts of the antenna booms had been painted black to reduce the surface area which could reflect light. A small midcourse correction was made approximately 44 hours after launch, and the initial high lunar orbit was established after 92.5 hours of

---

[45] Discussion with Dennis B. James, Bellcomm, Inc., July 25 and 28, 1969. The author and Mr. James studied photographs of Site A-3 and Frame M-100 and Mr. James pointed out the significance of these pictures to Mission II planning.

[46] Ibid. Compare Mission B Description document with that for Mission II.

cislunar transit time. The orbital parameters were: apolune, 1,850 kilometers; perilune, 196 kilometers. The Deep Space Network tracked Lunar Orbiter II for several days to obtain data for a more accurate analysis of the lunar gravitational effects on the spacecraft. After 33 orbits the spacecraft was transferred to the photographic orbit with a perilune of 49.7 kilometers.[47]

On November 18 Lunar Orbiter II commenced its photographic work. The photo subsystem performed well during all phases of the mission and covered each of 13 primary and 17 secondary sites as planned. Only Secondary Site II S-10.2 had to be rescheduled in the photographic plan, to avoid operating the spacecraft on batteries during photography, a procedure which would have violated a design restriction and resulted in a power shortage.

Several changes had been made in the photo subsystem of Lunar Orbiter II as a result of the first Orbiter mission:

1. The addition of an integrating circuit in the focal-plane-shutter control circuits to ensure that an output signal **represented a valid command pulse** (containing amplitude and duration) and **was** not caused by an electrical transient.

2. The addition of a filter on the 20-volt line to minimize electromagnetic interferences and possible triggering of photo subsystem circuits.

---

[47] Hall, TDS Final Report, Vol. III, Mission B Summary (No. 608-18), November 15, 1969, pp. 1-2, 1-3, 1-4.

3. The platen clamping spring tension was increased to ensure immobility of the film during exposure, improve film flatness, and maintain focus.

4. Reseau marks were pre-exposed on the spacecraft film in a specific pattern to assist in compensating for any non-linearities in the optical-mechanical scanner.[48]

The medium- and high-resolution photography was excellent in quality and indicated that the operation of the photo subsystem during exposure, processing, and readout was very good for the first portion of the film.

On November 20 Lunar Orbiter II photographed the impact point of Ranger VIII (Site II P-5).[49] On November 23 it recorded one of the most spectacular pictures of the lunar surface. The picture was taken as a result of the threat of Bimat stick and the need to move new film and Bimat onto the processor drum at regular intervals. A certain amount of the film would be wasted if no exposure were made and a choice arose as to the use of this "film-set" frame. One mission ground rule called for the frames to be used to take pictures of any areas in the Apollo zone of interest, should the spacecraft be over one at the time. On the other hand, Douglas Lloyd of Bellcomm,

---

[48] Lunar Orbiter II Photographic Mission Summary, NASA CR-883, prepared by the Boeing Company for Langley Research Center, October 1967, p. 33.

[49] Boeing Quarterly Technical Progress Report, October to December, 1966, p.5.

Inc., had suggested during mission planning that this particular "film-set" frame be used to take a photograph of the crater Copernicus when the spacecraft passed due south of it at a distance of 240 kilometers and a vertical altitude of 45 kilometers above the lunar surface. Twice his suggestion was turned down by NASA officials because of the Apollo ground rule. However, upon Lloyd's third suggestion Program officials consented, and the decision to make the picture came during actual mission operations.

The Lunar Orbiter's camera made a telephoto exposure through the 610 mm lens of the crater from a long, low, oblique angle to the lunar surface when lighting conditions were optimum for best contrast. The resultant picture revealed geographic and topographic features of the central portion of this 100-kilometer-wide crater which had never before been discerned. Dominating the center of the photographic frame were mountains rising over 300 meters from the crater floor. Behind them a ledge of bedrock and the crater's rim could be seen. Behind all of this the Gay-Lussac Promontory in the Carpathian Mountains towered 1,000 meters above the lunar surface on the horizon.

This and the oblique pictures of the Marius Hills and Reiner Gamma proved to be extremely valuable to the photogrammetrists, astrogeologists, and other scientists connected with the Lunar Orbiter and Apollo programs. The nation's news media described the Copernicus picture as "one of the great pictures of the century."[50]

<u>Lunar Orbiter II</u> ended its photographic acquisition on November 26, 1966, and flight controllers concluded the readout on December 7. Only one setback marred an otherwise unqualified success. The traveling-wave-tube amplifier (TWTA) failed on the final day of readout, and half of the photographs of secondary Site II S-1 were not obtained. This area was located at 41.1° east longitude and 3.2° north latitude in Mare Tranquillitatis.[51] However, priority readout of the wide-angle photo coverage of this site had previously been conducted, minimizing the seriousness of the loss.

The spacecraft's twenty micrometeoroid detectors recorded three impacts during nineteen days of the mission. These hits did not affect the performance of the spacecraft. <u>Lunar Orbiter I</u> had registered no hits, and program scientists believed that the <u>Lunar Orbiter II</u> hits may have been the result of the annual Leonid meteor shower.[52]

---

[50] Walter Sullivan, "Orbiter 2 Transmits Spectacular Close-ups of Moon," <u>New York Times</u>, December 1, 1966, p. 1. Douglas Lloyd's contribution to the planning of the Copernicus shot deserves recognition. His persistent belief that it could be done resulted in one of the program's outstanding photographic achievements. (Interview with Douglas Lloyd, Bellcomm, Inc., Washington, D.C., August 11, 1970.)

[51] <u>Lunar Orbiter II Photographic Mission Summary</u>, NASA CR-883, pp. 61, 86.

[52] Ibid., p. 86.

Lunar Orbiter II demonstrated its ability to obtain high-quality oblique photography of the near and far side of the Moon. It also obtained experimental convergent stereo telephoto pictures of one site, demonstrating the ability of the photographic subsystem to employ the stereo technique. Moreover, it showed that not all crater rays on the lunar surface were necessarily heavily cratered but that the Copernicus-Kepler region was unfit for landing sites. These achievements attested to the accuracy and precision with which the flight controllers were able to position the spacecraft for photographing specific objectives.[53]

Finally, the problem of overheating which had made more attitude control maneuvers necessary during the mission of the first Lunar Orbiter was overcome on the second mission. With the addition of a coating of S-13G paint, degradation of the thermal paint on the equipment deck of Lunar Orbiter II was substantially reduced. Thermal control of the spacecraft by planned thermal relief maneuvers was better integrated into the total flight operation plan for the second mission, and the spacecraft performance proved markedly better than that of the first Lunar Orbiter mission.[54]

---

[53] Ibid.
[54] Ibid.

The Third Orbiter Mission

The third mission differed slightly from the first two because it concentrated its photography on Apollo and Surveyor site confirmation instead of site search. To permit confirmation photography of sites both north and south of the lunar equator the spacecraft's orbital inclination was increased to 21°. The convergent stereo photography of Mission II had proved successful and potentially useful to the Apollo and Surveyor programs. It consisted of making two "footprints" of the same area on two successive orbits. To accomplish this at the higher orbital inclination, the camera would necessarily be tilted during one of the two sequences. Resolution of a convergent stereo picture pair was slightly degraded because of the camera tilt, and a loss of one-meter to two-and-one-half-meter, or perhaps three-meter, resolution resulted.[55]

The Air Force Aeronautical Chart and Information Center (ACIC) and the Army Map Service had evaluated the Mission II convergent stereo photography and had concluded that "this type of photography increases the topographic knowledge that can be obtained concerning potential landing sites."[56]

---

[55] Charles W. Shull and Lynn A. Schenk, U.S. Army TOPOCOM, "Mapping the Surveyor III Crater," *Photogrammetric Engineering*, Vol. XXXVI, No. 6, June 1970, pp. 547-554. This article gives a detailed analysis of how stereoscopic photography was utilized in site selection for *Surveyor III*.

[56] Lunar Orbiter Project Office, Langley Research Center, Lunar Orbiter Mission III Description, January 25, 1967, p. 15.

The Lunar Orbiter Project Office at Langley planned to include more convergent stereo coverage on Mission III as a result of the ACIC and Army Map Service (since January 1970, U.S. Army TOPOCOM) evaluations.

On November 15, 1966, a technical interchange meeting convened at the Jet Propulsion Laboratory to assess the various methods of calibrating the Lunar Orbiter's 610 mm high-resolution camera for the new photographic tasks. Precise geometric calibration was mandatory if stereo photography was to be conducted succesfully on the three remaining missions. The calibration was to be done at the photographic subsystem level, and the members of the meeting determined the method to use.[57] Leon J. Kosofsky coordinated the calibration activities.

Although primarily a reconnaissance photographic system, rather than a mapping system, the Lunar Orbiter photo subsystem was upgraded after Mission I. The Aeronautical Chart and Information Center and the Army Map Service had previously argued that the use of reseau marks on the camera film or a grid on the camera lens would greatly facilitate the utilization of photographic data for purposes of lunar mapping. Langley accepted the idea of pre-exposing reseau marks on the camera film for Mission II and all subsequent

---

[57] Memorandum from Lee R. Scherer to Clifford Nelson, Langley Research Center, Subject: Geometric Calibration of High Resolution Camera for Mission C, December 20, 1966.

missions.

On January 5 the photo subsystem for Spacecraft 6 (the third flight spacecraft) was installed, and Boeing conducted the functional check-out with the Deep Space Instrumentation Facility. The spacecraft's inertial reference unit (IRU) was taken out, tested, and reinstalled and the actuator for solar panel 3 was replaced. Retesting at Hangar S was accomplished by January 13 in preparation for mating with the launch vehicle.[58]

Meanwhile, on January 5 the Ad Hoc Surveyor/Orbiter Utilization Committee of OSSA had approved the plan for the third Lunar Orbiter mission:

> Mission III is primarily designed to photograph promising areas that have been identified by screening Lunar Orbiter I and II photographs and for which additional data is needed to confirm their adequacy as Apollo and/or Surveyor landing sites. In addition Mission III will provide photography of broad scientific interest as did Missions I and II.[59]

The mission would also obtain precision trajectory information to be used in improving the definition of the lunar gravitational field and measurements of micrometeoroid flux and of radiation dosage levels in near-lunar environment for use in evaluating the spacecraft's performance.

---

[58] *Project Lunar Orbiter, Narrative Analysis*, Langley Research Center, January 17, 1967.

[59] Lunar Orbiter Mission III Description, p. 1.

Finally <u>Lunar Orbiter III</u> would serve as a target for the Manned Space Flight Tracking Network and the Orbit Determination Program.[60]

The Launch Readiness Review for <u>Lunar Orbiter III</u> and for the back-up (Spacecraft 7) was held at the Eastern Test Range facilities on January 17. Both Orbiters were found to be ready for launch, and personnel working with Spacecraft 6 proceeded with the preparations for that event. The tentative date for launch was February 3.[61]

Boeing and Eastman Kodak were attempting to resolve the problems which had caused minor film processing defects on the first two missions. Manufacturing irregularities and bubbles in the Bimat had been the chief causes of these defects. As it turned out, localized Bimat processing defects continued to appear on some photographs from all five missions, despite attempts to correct the condition. Still unresolved as the third launch approached was the failure of the TWTA aboard <u>Lunar Orbiter II</u>. However, Boeing engineers were modifying this component so that excess heat build-up could be removed during the flight, thus prolonging the tube's lifetime. Readout times would also be reduced in the event of a heat build-up, and flight controllers would

---

[60] Lunar Orbiter C Mission Objectives, unsigned memorandum, January 24, 1967.

[61] <u>Project Lunar Orbiter, Narrative Analysis</u>, Langley Research Center, February 15, 1967.

monitor the flow of electrical current through the traveling-wave-tube amplifier, since program scientists considered any irregularities in the flow to be an indication of pending trouble in it.[62]

Lunar Orbiter III lifted off of Pad 13 at the Eastern Test Range at 01:17 Greenwich Mean Time on February 5, 1967. (The February 3 launch window had been canceled because of problems encountered in the ground power-supply system at Launch Complex 13.) Despite numerous pre-launch problems the liftoff was successfully accomplished on a flight azimuth of 80.8° at the start of the February 5 launch window. Ground control placed the Agena-spacecraft combination in a parking orbit for approximately ten minutes before injecting it into a cislunar trajectory.[63]

Following injection the spacecraft separated from the Agena, deployed its solar panels and antennas, and acquired the Sun as an attitude reference. Seven hours into the mission, flight controllers commanded Lunar Orbiter III to turn on its Canopus star tracker and give a star map before Canopus acquisition. It executed this command successfully. On Monday, February 6, at 37 hours into the mission

---

[62] Memorandum from SL/Manager, Lunar Orbiter Program, to SE/Deputy Associate Administrator for Space Science and Applications (Engineering), January 24, 1967.

[63] Hall, TDS Final Report, Vol. IV, Mission C Summary (No. 608-19), March 1, 1969, p. 1-2.

the Space Flight Operations Facility tracking <u>Lunar Orbiter III</u> commanded a midcourse correction maneuver to adjust the spacecraft's cislunar trajectory in order to hit the preplanned aiming point for deboost into lunar orbit. As on previous missions, the midcourse maneuver was so accurate that no second maneuver was required.[64]

At 4:54 p.m. Eastern Standard Time on February 8 <u>Lunar Orbiter III</u> fired its 100-pound-thrust rocket engine for 9 minutes, 2.5 seconds to decelerate the spacecraft into its initial orbit. The parameters were: apolune, 1,801.9 kilometers; perilune, 210.2 kilometers; inclination, 20.93°; period of orbit, 3 hours 25 minutes.[65] Ground control tracked the spacecraft in the initial orbit for approximately four days (25 orbits) to obtain data for analysis of the lunar gravitational effect. Following this the spacecraft was transferred to a new orbit with a low perilune of 55 kilometers and an apolune of 1,847 kilometers.[66] Inclination to the lunar equator was 20.9°[67]

As <u>Lunar Orbiter III</u> had executed its deboost maneuver <u>Lunar Orbiter II</u> was still in orbit around the Moon. On February 6 ground control began tracking both spacecraft

---

[64] Boeing Quarterly Technical Progress Report, January to March 1967, p.4. See also Status of Lunar Orbiter III (as of 8 a.m. EST), February 7, 1967.

[65] Status of Lunar Orbiter III, February 9, 1967, p. 3.

[66] Hall, <u>TDS Final Report</u>, IV, pp. 1-2, 1-3, 1-4.

[67] Boeing Quarterly Progress Report, January to March 1967, p. 4.

simultaneously, thus demonstrating its ability to track two spacecraft in different orbits around the Moon at the same time. This exercise greatly extended the usefulness of each mission by providing simultaneous telemetry on the two orbiting spacecraft. Monitoring showed that all Lunar Orbiter II subsystems were operating normally.[68]

Lunar Orbiter III began its photographic mission on February 15 over primary Site III P-1 at 35°15" east longitude, 2°55" north latitude, near the crater Maskelyne F in the southeastern region of Mare Tranquillitatis. The first readout in the primary mode revealed photographs of excellent quality. A solar flare occurred at 12:54 p.m. EST on February 13. Though it had a high amount of optical activity, there was little of the proton activity that could have presented a danger to the film on board the spacecraft.[69] The first readout revealed no fogging of the film and indicated that all subsystems were working normally.

The film advance mechanism in the readout section of the photo subsystem of Lunar Orbiter III began to show erratic behavior even during the mission's photographic phase. Because of this, program officials decided to begin final readout earlier than planned. Ground control at the

---

[68] Status of Lunar Orbiter III, February 9, 1967, p. 4.

[69] Status of Lunar Orbiter III (as of 3:30 p.m. EST), February 13, 1967; and Status of Lunar Orbiter III, February 16, 1967.

DSN decided not to photograph secondary Site S-32, an oblique shot of the Grimaldi crater area. A total of 211 out of 212 planned frames had been exposed when, at 1:36 a.m. EST on February 23, flight controllers commanded the spacecraft to cut the Bimat, closing out the photographic portion of the third mission. By March 1, readout had been completed for 114 frames of photography, or 54% of the total. Film advance through the readout gate was intermittently hampered during this time, but no no photography was lost.[70]

Then suddenly on March 4 readout ceased. Of the 211 frames, 72 still remained to be read out, but the worst had happened. The film advance motor had burned out, and the 72 frames remained on the take-up reel. Program engineers concluded that an inexplicable electrical transient had scrambled the photo system's logic, causing the motor to run out of control. Nonetheless, 75% of the photographic data had been transmitted to Earth before this failure. The decision to begin readout earlier than planned had proved very prudent indeed.[71]

Mission III photography displayed the finest overall quality thus far obtained in the program. The quality was due ir

---

[70] Status of Lunar Orbiter III, February 23, 1967; and March 1, 1967.

[71] Boeing Quarterly Technical Progress Report, January to March 1967, p. 4. See also Lunar Orbiter III Photography, NASA CR-984, prepared by the Boeing Company for Langley Research Center, February 1968, p. 108, for a detailed report of the failure.

part to the use of more diversified photographic procedures, including the use of precisely oriented camera axis over a wide range of tilt angles and azimuth. The exposure sequencing modes were varied and used more extensively. Relaxation of earlier photographic constraints, higher orbit inclination, and extended stereoscopic photography resulted in greater coverage over a wider range of latitude and successful photography under extreme illumination conditions.[72]

Among other important sites <u>Lunar Orbiter III</u> photographed the <u>Surveyor I</u> landing area, permitting the location and identification of the spacecraft on the Moon's surface in Telephoto Frame 194 of Site III P12a.[73] This and other accomplishments proved the reliability, accuracy, and versatility of the spacecraft in its lunar exploration mission and gave program officials the confidence to attempt more complex precision photography on the two remaining missions.

---

[72] <u>Lunar Orbiter II Photography</u>, NASA CR-984, p. 120.

[73] <u>Ibid.</u>

CHAPTER X

MISSIONS IV AND V: THE LUNAR SURFACE EXPLORED

The first three missions essentially satisfied the Apollo requirements for photographic data of potential landing sites. This opened the two remaining missions to other work. Photography could concentrate on specific areas of the Moon which scientists from various disciplines wished to explore more closely. It could also enable NASA cartographers to compile a much more nearly complete lunar atlas than any then in existence.

Preparing for the Fourth Mission

As approved by the Ad Hoc Surveyor/Orbiter Utilization Committee on May 3, 1967, Mission IV would attempt to accomplish some of the objectives not directed towards fulfilling Apollo needs. Specifically it would "perform a broad systematic photographic survey of lunar surface features in order to increase the scientific knowledge of their nature, origin, and processes, and to serve as a basis for selecting sites for more detailed scientific study by subsequent orbital and landing missions."[1]

This mission, unlike the first three, required that Lunar Orbiter IV fly a nearly polar orbit. In such an orbit

---

[1] Lunar Orbiter Project Office, Langley Research Center, Lunar Orbiter Project Mission IV Description, April 26, 1967, p. 3.

the spacecraft would acquire contiguous photographic coverage of a minimum of 80% of the front side at 50 to 100 meters resolution. It would photograph as much of the Moon's far side as possible at the best possible resolution. The spacecraft's photographic subsystem would carry enough film for 212 frames, and ground control planned to read out all photography in the priority mode immediately after processing as a precaution against any mechanical failure in the subsystem. A final readout would be available if necessary.[2]

In preparation for the fourth mission the Lunar Orbiter Project and Program Offices conducted a flight readiness review on April 13, 1967. On March 13, Spacecraft 7 (the fourth flight spacecraft, or Lunar Orbiter IV) had been removed from storage at the Kennedy Space Center to begin Hangar S integration and checkout tests. Launch readiness was scheduled for May 4, and no problems were encountered during the Hangar S activities.[3] The flight readiness review found Lunar Orbiter IV and the backup (Spacecraft 3) ready for launch.[4]

Because the fourth Orbiter would fly a high polar orbit, it would be exposed to the Sun almost the entire

---

[2] Ibid., p. 4.
[3] Project Lunar Orbiter, Narrative Analysis, Langley Research Center, March 15, 1967, and April 17, 1967.
[4] Memorandum from SL/Manager, Lunar Orbiter Program, to SE/Deputy Associate Administrator for Space Science and Applications (Engineering), April 14, 1967, pp. 2-3.

mission, necessitating certain changes on the spacecraft. A modified charge-controller component was installed to reduce the rate of charge in the power system. Boeing engineers covered about 20% of the exterior of the equipment deck with mirrors to increase its heat rejection capability. A damaged micrometeoroid detector was removed and another unit installed. Finally the Inertial Reference Unit was removed for replacement of a failed capacitor. After reinstallation it successfully completed two attitude control system tests.[5]

During the weeks before the fourth launch the Program Manager showed some concern over the failure of NASA's Applications Technology Satellite (ATS II) to achieve its planned circular orbit around the Earth on April 6.[6] NASA officials attributed the improper orbit to failure of the Agena rocket to reignite in orbit. Unofficially ATS program management said the cause for the reignition failure was failure of the Agena's Propellant Isolation Valve (PIV) to close after the first burn. Scherer hoped the PIV for the Lunar Orbiter IV Agena would test out successfully before April 27, the planned date for the mating of the Agena with the Atlas

---

[5] Ibid.

[6] NASA, Executive Secretariat, Program and Special Reports Division, Space Flight Record, 1958-1968, December 31, 1968, p. 25.

booster.[7] Lewis Research Center personnel responsible for the Agena took corrective actions and installed a reworked valve in time for the launch. The reinstallation took less than one month to complete, and it did not jeopardize the launch date.

Two areas involving previous mission and ground test problems also pertained to the successful performance of the fourth and fifth missions. The traveling-wave-tube amplifier aboard Lunar Orbiter II had experienced high helix current. Ultimately it had failed to turn on during the final readout phase, and some data were lost. The TWTA onboard Lunar Orbiter III had also experienced overheating from high helix current and power output variations from temperature changes. Worse yet, the TWTA in the ground spacecraft for the Mission D Simulation Test failed to perform successfully under mission conditions. The component was undergoing close examinations to determine the mode of failure. A delay of the fourth mission would hinge upon the seriousness of the test findings and the difficulty in resolving the problem.[8]

Failure in the photographic subsystem presented the other area of questionable spacecraft performance. Readout

---

[7] Memorandum, SL/Manager to SE/Deputy Associate Administrator, p. 1.
[8] Ibid., p. 2.

problems had marred the success of Lunar Orbiter III with unwanted repetition in readout and the inability of the film transport system to move film. Program investigators had not pinpointed the causes of these failures. However, the ten-day Mission D Simulation Test, just completed on April 12, partially compensated for these failures. During the test no problems involving readout had occurred, increasing the likelihood of a successful fourth mission.

The Fourth Orbiter Mission

Last minute tests did not reveal any problems of a magnitude serious enough to delay a launch, and on May 4 Lunar Orbiter IV rode into space atop its Atlas-Agena D launch vehicle at 18:25 Eastern Daylight Time (EDT) from Launch Complex 13 at Cape Kennedy on an azimuth of 100.8°. About thirty minutes after liftoff the Agena injected the spacecraft into a cislunar trajectory. Early tracking data indicated that it was on course, and the first midcourse maneuver was scheduled for 13:00 EDT on May 5.[9]

Early in Lunar Orbiter IV's journey to the Moon the Canopus star tracker experienced difficulty acquiring Canopus. Glint from the Sun and earthshine probably were

---

[9] Lunar Orbiter Program Office, NASA, Post Launch Mission Operation Report (MOR) No. S-814-66-04, Lunar Orbiter IV Post Launch Report #1, May 5, 1967.

the causes of this trouble. The star tracker did lock onto
a celestial body, but flight controllers were not sure if it
had acquired Canopus or the planet Jupiter, which was also in
its field of view. Program operators planned to correct this
situation by staging a roll reference maneuver during the
first midcourse correction.[10]

Passing through the Van Allen Belt, Lunar Orbiter IV
experienced a higher dose of radiation than had the previous
Orbiters: 5.5 rads recorded by the radiation dosimeter for
the film supply cassette, versus 0.75 rads on earlier Orbiters.
However, the dosimeter for the camera storage loopers
registered 0.0 rads when it was turned on after the
spacecraft had traversed the Van Allen Belt.[11]

Shortly after noon EDT on May 5 Lunar Orbiter IV
executed the planned midcourse maneuver to line the spacecraft up with the aiming point before deboost into orbit
around the Moon. At 11:08 EDT on May 8 the spacecraft's
rocket burn deboosted the Orbiter into an initial near-
polar orbit around the Moon, with 6,111-kilometer apolune,
2,706-kilometer perilune, 85.5° inclination to the lunar
equator, and 12.01-hour period of orbit.[12]

---

[10] Ibid.
[11] Ibid., p. 2.
[12] Ibid., Lunar Orbiter IV Post Launch Report #3, May 9, 1967.

All subsystems performed well and within acceptable temperature limits up to this point. Flight controllers at the Deep Space Network facilities commanded the spacecraft to scan the Goldstone Test Film at 7:30 p.m. EDT on May 9 in order to check the readout and communications subsystem. The DSN stations at Goldstone, California, and Woomera, Australia, read out the film and received data of excellent quality. The TWTA onboard the spacecraft had been turned on for readout and would remain on for the duration of the mission. The spacecraft would execute thermal control maneuvers to suppress any overheating tendency of the TWTA during the mission. Readings of the radiation dosimeters for the film storage cassette continued to stand at 5.5 rads, while the dosimeter for the storage loopers indicated a change from 0.0 to 0.5 rads. Ground control attributed this to background radiation from space, which did not threaten the film.[13]

In its sixth orbit around the Moon Lunar Orbiter IV began its first photographic pass at 11:46 a.m. EDT on May 11. As the spacecraft sped from south to north the photo subsystem exposed five sets of four frames each at intervals ranging from 30 to 40 minutes. At the high altitude, image-motion compensation did not enter into the photographic

---

[13] Ibid., Lunar Orbiter Post Launch Report #4, May 11, 1967.

process. Passing over the vicinity of the lunar north pole, the spacecraft dropped out of sight and radio contact with Earth. How could it conduct farside photography without direct communication with flight controllers? The key to the Orbiter IV farside photography as well as to all farside photography of the five Lunar Orbiter missions was the Flight Programmer, previously discussed.

Originally Boeing had designed the Programmer for a command storage capacity of sixteen hours, twice the length of time in which any of the DSN ground receiving stations would be out of line-of-sight communications with the spacecraft. This represented a safety margin of eight hours, should one of the stations fail to acquire the spacecraft. The storage capacity mean that flight programmers could store commands to be executed up to sixteen hours following storage without any further command from Earth. Thus, during the periods when the spacecraft was out of sight of the Earth, it was already programmed to conduct photography of the lunar far side.[14]

Heading south from the north pole Lunar Orbiter IV took one frame of the Moon's far side as it reached apolune (6,111.3 kilometers). By 8:40 p.m. EDT May 11, it had exposed a total of 27 frames, and flight controllers commanded the readout of this photography to begin. The

---

[14] Costello interview.

first high- and medium-resolution pictures turned out excellently.[15]

Despite this apparent success, the spacecraft had already developed a serious problem which threatened to jeopardize the whole mission. Telemetry data indicated that after the second set of four frames had been exposed, the camera thermal door failed to close until ground control had sent additional commands to close it. After the third set of four frames had been made, spacecraft telemetry did not confirm if the door had opened sufficiently. Flight controllers initiated a preliminary corrective action by commanding the door to open far enough in advance of the fourth set's exposure time to allow for additional commands if required.

NASA and Boeing engineers began immediately to analyze the problem. The danger of the thermal door's failing in the closed position and making all further photography impossible forced flight controllers to fly the spacecraft with the door open. The open door created a danger of light leakage, which could fog portions of the film. Flight controllers had to strike a delicate balance between prohibiting light leaks and preventing the temperature within the subsystem from dropping below the dew point of the gas

---

[15] Post Launch MOR S-814-66-04, Lunar Orbiter IV Post Launch Report #6, May 12, 1967.

which pressurized it. Too low a temperature could cause moisture condensation on the camera lens window and thus reduce the contrast and resolution of the photographs. Maintaining a balance between these two conditions led to extra attitude control maneuvers.[16]

The danger of light leakage revealed itself early on May 13 during the readout of the exposures which the spacecraft had made since ground control had initiated contingency measures to cope with the camera thermal door problem. Portions of the photographs were light struck. NASA engineers deduced the mishap by comparing readout results of film that had been kept in the spacecraft's camera storage looper for one half hour with film that had been there five hours and longer. The quality of the exposures declined with the length of time the film had been in the looper before readout.[17]

Lunar Orbiter Program personnel from Langley, Boeing, and Eastman Kodak attempted to solve the problem of the door. Flight controllers devised and executed several tests to assess its reliability. These showed that the door could be partially closed, then reopened. Further tests placed the spacecraft in several orientations to the Sun with the door

---

[16] Ibid.

[17] Ibid., Lunar Orbiter IV Post Launch Report #7, May 15, 1967.

partially closed. Ground control monitored the thermal response of the camera lens window and commanded the spacecraft to take photographs. On May 16 these photographs were read out, and they indicated that light leaks had ceased. Program officials concluded that their procedures were effective. However, the low contrast of some pictures indicated probable fogging of the lens window due to moisture condensation at lower temperatures. Ground control maneuvered the spacecraft to raise the temperature of the lens window on orbit 14 and subsequent orbits.[18]

As of May 19 Scherer could report to NASA Administrator James E. Webb that the Langley/Boeing flight operations team had the photographic fogging problem under control. The team had established the following subjective grading system for *Orbiter IV* pictures: 1) excellent quality, 2) light fogging, 3) heavy fogging, and 4) blank. The most recent high-resolution photographs fell into the first or second categories, with most being graded excellent. A preliminary analysis of the photographic coverage during the first $60°$ of lunar longitude arc indicated that 64% of this area had been covered by grade 1 or 2 photography.[19]

Early on Saturday morning, May 20, ground control

---

[18] *Ibid.*, Lunar Orbiter IV Post Launch Report #8, May 17, 1967.

[19] *Ibid.*, #9, May 22, 1967

picked up an anomaly during readout. The readout drive mechanism turned off in a normal manner without being commanded to do s   Ground control restarted it, but after scanning a short segment of film it stopped abruptly. Throughout the day this start-stop situation repeated itself; the distance scanned varied from 5 to 30 centimeters. Langley and Boeing engineers suspected the readout encoder was falsely indicating a full readout looper. They began to analyze the problem while primary readout proceeded. Pictures obtained through readout proved that the new operational procedures for the camera thermal door continued to be effective, and no change in photography schedules was necessary at that time.[20]

By 8:00 a.m. EDT on May 25 <u>Lunar Orbiter IV</u> was in its thirty-fourth orbit around the Moon and had photographed its surface as far as the 100° west meridian. Ground control had recovered photographs up to about the 75° west meridian. The sector from 90° east to 45° east meridian, which the Orbiter had first photographed, had been photographed again from apolune because fogging had degraded the quality of the perilune pictures. While photography proceeded well, flight controllers believed that they had brought the premature

---

[20] <u>Ibid.</u>, Lunar Orbiter Post Launch Report #10, May 22, 1967.

termination of readout under control. They used a repetitive series of commands to prevent the noisy encoder from stopping readout until commanded to do so.[21]

Between May 21 and May 25, while problems with the thermal door and the readout encoder were being resolved, Lunar Orbiter IV experienced increased radiation dosage from solar flare particle events. Trutz Foelsche, primary investigator for the Lunar Orbiter radiation experiment, was able to make preliminary conclusions about the potential hazards to Lunar Orbiter IV based upon early data which the Space Flight Operations Facility had obtained from the spacecraft's two dosimeters. On May 21 a solar particle event had produced low-energy protons whose energy levels did not exceed 20 Mev. Since they had little energy these protons would hardly affect the camera film. Moreover, he concluded, the May 21 event was much less serious than the event of September 2, 1966, which Lunar Orbiter II had encountered, and the Orbiter had experienced no film fogging.[22]

---

[21] Ibid., Lunar Orbiter IV Post Launch Report #11, May 25, 1967.

[22] Memorandum from Martin J. Swetnick, SL/Scientist, to File, June 1, 1967, Subject: Status of assessment of Lunar Orbiter IV radiation detector data. See also: Trutz Foelsche, "Radiation Measurements in LO I - V (Period August 10, 1966—January 30, 1968)," Langley Research Center, for a detailed analysis of the data on radiation doses returned to Earth by the five Lunar Orbiter spacecraft.

On the thirty-fifth orbit around the Moon <u>Lunar Orbiter IV</u> experienced worsening readout difficulties. These brought a quick decision to cut the Bimat to escape the high probability that the Bimat would stick to the film, thus ending the photographic mission. At this time the photographic subsystem had exposed and processed 163 frames. Ground control successfully commanded <u>Lunar Orbiter IV</u> to cut the Bimat, but final readout presented more problems.[23]

The erroneous encoder signals hindered film transport from the take-up spool considerably, and ground control had to improvise a non-standard procedure to get around this condition. Sending false picture-taking commands, mission controllers inched the film towards the take-up spool and then moved short segments of film back through the readout gate. Using this procedure they successfully recovered 13 additional frames at the end of the film which might otherwise have remained between the processor and the readout looper. Then ground control sent commands to the spacecraft to apply tension throughout the film system. Following this the system responded normally to readout operations. Only 30 of the 163 frames which had been exposed remained to be recovered. NASA ground stations completed final readout on June 1.[24]

---

[23] Post Launch MOR S-814-66-04, Lunar Orbiter IV Post Launch Report #12, May 29, 1967.
[24] <u>Ibid</u>.

Lunar Orbiter IV photography had covered 99% of the Moon's near side at a resolution exceeding by ten times the best Earth-based telescopic photography. This coverage revealed significant, heretofore unknown, geological detail in the polar and limb regions of the Moon. Unofficially the Orbiter IV photography increased to 80% the coverage of the far side of the Moon obtained during the first four Orbiter missions. These accomplishments attested to the high degree of organization in the flight operations of the fourth mission in the face of the problems that had been encountered.[25]

Its photographic mission ended, Lunar Orbiter IV proceeded into its extended mission. Program officials planned to change the spacecraft's orbit so that it would approximate that planned for Lunar Orbiter V. The additional information which ground control could obtain about the Moon's gravitational environment by tracking Lunar Orbiter IV and analyzing the telemetry data would prove valuable in planning the final Orbiter mission. In addition ground stations continued to track the second and third Orbiters. Lunar Orbiter II, launched in November 1966, was moving

---

[25] Ibid., Lunar Orbiter IV Post Launch Report #13, June 5, 1967. The U.S. Air Force Aeronautical Chart and Information Center subsequently determined that of the total farside coverage of the Moon only 60% was usable for purposes of mapping (confirmed in a telephone conversation with Leon J. Kosofsky, Lunar Orbiter program engineer, September 15, 1967).

closer to the Moon's surface on an inevitable collision course. Program officials planned to raise its orbit, thus extending its lifetime. Lunar Orbiter III would undergo a plane change in its orbit in addition to having it raised. The change would provide new data on the lunar gravitational field for use in further mission planning and in the Apollo Program.[26]

Preparations for the Fifth Mission

In March 1967, before the fourth mission, a working group within the Lunar Orbiter Program developed tentative objectives for the fifth and final mission. These called for a multi-site scientific mission with the capability of reexamining the eastern Apollo sites. A subgroup formed to determine specific target sites for the photographic mission of the last flight. As in the past the Lunar Orbiter Project Office at Langley coordinated all mission planning activities.[27] On March 21 the entire working group met at Langley to review the preliminary plans. The results of the review were sent to Boeing for further consideration before a presentation to the Ad Hoc Surveyor/Orbiter Utilization

---

[26] Ibid.
[27] Memorandum from SL/Manager, Lunar Orbiter Program, to SL/Director, Lunar and Planetary Programs, Subject: Lunar Orbiter Mission 5 Planning, March 9, 1967. See also Minutes of the March 7, 1967, meeting of the Mission V Planning Group, NASA Headquarters.

Committee at the end of the month.

The Lunar Orbiter Mission V Planning Group, which had come into being in March, met at the Jet Propulsion Laboratory on May 26 to review the Boeing Company's preliminary mission design for the fifth Orbiter. Of special interest was the problem of orbit design. The Group worked out an orbit design which would meet the needs of the multi-site mission without violating spacecraft design restrictions. The orbit would have an inclination of $85°$ to the Moon's equator. The perilune altitude would be low enough to allow two-meter-resolution photography on vertical photographs instead of one-meter, in order to obtain more useful convergent stereo photography at the higher altitude of 100 kilometers. At the higher perilune the cross-camera tilt would be reduced, offering better resolution on the convergent stereo photographs. At the same time, increasing the perilune altitude broadened the coverage of the science sites.[28]

The Planning Group decided to keep the Lunar Orbiter V apolune as low as possible and no higher than 1,500 kilometers above the Moon. Lighting angles from the morning terminator would range from $8°$ to $24°$--angles offering the greatest potential relief rendition of surface features to

---

[28]Minutes of the May 26, 1967, meeting of the Mission V Planning Group, p. 2.

assist scientists in analyzing topographic and geologic aspects of the lunar surface.[29]

By June 14 the Lunar Orbiter Program Office had the completed plan for the fifth mission, and the Ad Hoc Surveyor/Orbiter Utilization Committee approved it on the same day. As a result of the review of <u>Lunar Orbiter IV</u> photography, mission planners at Langley changed almost 50% of the sites they had initially selected for the fifth mission.[30]

Lunar Orbiter V Mission Objectives

The fifth mission's objectives can be divided into two categories: photographic and non-photographic. The former composed the primary part of the mission, the latter the secondary. The spacecraft would perform five basic photographic tasks. Task 1 entailed additional Apollo landing site photography, employing three modes of photography: near-vertical, convergent telephoto stereo, and oblique. Task 2 would accomplish broad survey photography of unphotographed areas on the Moon's farside. Task 3 was to take photos of additional Surveyor landing sites of

---

[29] Ibid.

[30] Lunar Orbiter Mission V Description approved by the Ad Hoc Surveyor/Orbiter Utilization Committee on June 14, 1967, prepared by the Lunar Orbiter Project Office, Langley Research Center, July 8, 1967, pp. 2-3.

high scientific interest to investigators. Task 4 would have the spacecraft concentrate on potential landing sites for later Apollo Program missions, with particular stress on their scientific value. Finally, Task 5 was related to the fourth in that it encompassed photography of a wide range of scientifically interesting sites.[31]

The second category of mission objectives did not differ markedly from the first four missions. It included the following: 1) acquisition of precision trajectory information for use in improving the definition of the lunar gravitational field; 2) measurement of the micrometeoroid flux and radiation dose in lunar environment, primarily for analysis of the spacecraft's performance; 3) provision to the Manned Space Flight Network tracking stations of a spacecraft which they could track for purposes of evaluating the network and the Apollo Orbit Determination Program.[32]

<u>Lunar Orbiter V</u> would fly a nearly polar orbit inclined 85° to the Moon's equator. The spacecraft would deboost into an initial orbit with an apolune of 6,000

---

[31] <u>Ibid.</u>, pp. 4-7. The responsibilities for follow-on lunar exploration were assigned to the Apollo Program and were under the Apollo Lunar Exploration Program. This program differed from the Apollo Applications Program, which was concerned with Earth-orbit applications of Apollo hardware and technology.

[32] Lunar Orbiter Mission V Description.

kilometers and a perilune of 200 kilometers. In this orbit it would take photographs of the lunar far side. Finally, the spacecraft would maneuver to a new orbit with an apolune of 1,500 kilometers and a perilune of 100 kilometers to execute the remainder of the photographic tasks.[33]

As approved the mission plan called for a total of 212 film frames to be exposed. Of these it had allocated 44 frames to Apollo tasks and 168 frames to scientific areas, including those thought suitable for the later Apollo missions and for Surveyor landing sites. Five Apollo sites along the equatorial zone, ranging from 42°56' east longitude to 36° 11' west longitude and from 0°45' north latitude to 3°30' south latitude, would be photographed. Potential Apollo Program sites which Lunar Orbiter V would photograph included: the Littrow rilles; the Sulpicius Gallus rilles; the Imbrium flows; the craters Copernicus, Dionysus, Alphonsus, Dawes, and Fra Mauro; Copernicus secondary craters; the domes near Gruithuisen and Gruithuisen K; the Tobias Mayer dome; the Marius hills; the Aristrachus plateau; the area of Copernicus CD; and the areas south of the crater Alexander on the northern edge of Mare Serenitatis.[34]

What did mission planners use as criteria for selecting science sites? Donald E. Wilhelms of the United

---

[33]Ibid., pp.11-13.
[34]Ibid., pp. 18-21.

States Geological Survey, working with the Lunar Orbiter Program Office, described one of the major criteria:

> The primary criterion for selection of Mission V sites was freshness of the features in the site. Earlier Orbiter missions have shown emphatically that most lunar terrain has a subdued appearance at all Orbiter scales so that little new is learned from high resolution photography. Fresh young craters (mostly light) and fresh young rock units (mostly dark) that are not yet much modified by repeated cratering and wasting potentially reveal the most about rock type and origin, both in photographs and when sampled on the ground. Old terrains show effects of the processes that waste lunar slopes, and though these are of interest, they seem to be sufficiently sampled in high resolution photography by earlier Orbiter missions, except for very high and steep slopes. A few high and steep slopes and other non-fresh targets have been selected for the purpose of rounding out terrain sampling.[35]

The fifth Orbiter mission would perform the most exacting, precision photography of all five missions. It also had the experience of the previous four flights to call upon in establishing greater confidence in mission controllers concerning operational procedures. As a result they could demand more of Lunar Orbiter V. Nevertheless the spacecraft exhibited several problems during preflight tests and check-out at Cape Kennedy. The most serious problem

---

[35] Ibid., p. 22. Wilhelms subsequently described each site which Lunar Orbiter V would photograph, giving its geographic location and the main features of scientific interest. Lunar Orbiter photographs of each site accompanied his descriptions. Mission IV photography proved extremely helpful in refining estimates of site freshness, in relocating Mission V sites, and in rejecting some previously selected sites.

developed when the bladders of the oxidizer tanks began to leak. The leaks forced NASA to return Spacecraft 3 (the fifth flight spacecraft) to Boeing in Seattle on May 12. It arrived there on May 17 and the oxidizer bladders were replaced by June 6. It was then returned to Hangar S at Cape Kennedy on June 16 for retesting. Integration and checkout with the launch vehicle took place on July 12, with final mating on July 19.[36]

By July 27 Lunar Orbiter V had successfully completed pre-launch tests and had been mated with the launch vehicle in preparation for an August 1 launch.[37] Program officials subsequently conducted a simulated launch exercise on July 28. The fifth mission was about to begin.

The Final Mission

A NASA Boeing Lockheed team launched Lunar Orbiter V successfully from Launch Complex 13 at Cape Kennedy on August 1, 1967, less than one year after the first Orbiter had made its long journey to the Moon. The countdown proceeded smoothly throughout the day with only one anomaly in the Agena, causing a short hold. Then it resumed until mid-afternoon. The launch was scheduled for 4:09 p.m. EDT,

---

[36] Project Lunar Orbiter, Narrative Analysis, Langley Research Center, June 13, 1967 and July 18, 1967.

[37] Status of Lunar Orbiter E, July 27, 1967.

but a rain storm delayed it for two and one half hours. The threat of postponing the launch grew serious because the launch window on August 1 lasted only from 4:09 p.m. to 8:00 p.m. EDT. The threat was significant to the mission because, if the weather forced a delay until the launch window of the following day, a partial loss of farside photography would result. Lunar Orbiter V was targeted for a high, elliptical polar orbit so that it could perform photography over the Moon's entire surface. The Moon rotates 13° of arc on its axis per Earth-day. A delayed launch of one day would mean the loss of a 13° portion of the lunar far side to darkness.[38]

Fortunately the weather improved, and the countdown resumed. Launch control fired the Atlas-Agena carrying Lunar Orbiter V on its way to the Moon at 6:33 p.m. EDT. In the monitoring room program officials sat watching the large display panels as various signals lit up, telling them that the different marks of the launch operation had been achieved. Early telemetry data indicated that all systems were functioning excellently. Fifty minutes into the mission the Deep Space Tracking Network station at Woomera, Australia, acquired radio contact with the spacecraft. It confirmed for

---

[38] Interview with A. Thomas Young, Lunar Orbiter Project Office, Langley Research Center, obtained during launch operations at Cape Kennedy, August 1, 1967.

ground control that the spacecraft had separated from the Agena and deployed its solar panels and two antennas and that its power system was operating on solar energy. All subsystems continued to perform normally and within acceptable temperature limits.[39]

Flight controllers at the Jet Propulsion Laboratory, where DSN operations shifted after the launch, executed the first midcourse maneuver at 2 a.m. EDT on August 3. This corrected the spacecraft's trajectory, which was about 7,000 kilometers off the aim point, for the deboosting maneuver into lunar orbit. *Lunar Orbiter V* carried out a roll maneuver of +42.1°, a pitch maneuver of +29.1° and a burn of its velocity control engine of 26 seconds. The resulting velocity increment of 29.76 meters per second was sufficient to put the spacecraft on course for arrival at the planned aiming point at the specified time. No second midcourse correction was necessary.[40]

During the cislunar transit the spacecraft had no difficulty acquiring Canopus before the midcourse maneuver.

---

[39] Lunar Orbiter Project Office, Langley Research Center, Lunar Orbiter Project Mission Countdown Document LOTD-106-4, approved July 5, 1967. The document lists every command and milestone in the network countdown procedure, beginning at T minus 505 minutes. See also Lunar Orbiter Program Office, NASA, Post Launch Mission Operation Report (MOR) No. S-814-67-07, Lunar Orbiter V Post Launch Report #1, August 2, 1967.

[40] Post Launch MOR S-814-67-07, Lunar Orbiter V Post Launch Report #2, August 3, 1967.

The radiation dosimeter at the film supply cassette registered a dose of 0.75 rads as the spacecraft passed through the Van Allen Belt. After transit the dosimeter in the camera storage looper was turned on, and it registered 0.0 rads. The ship recorded no micrometeoroid hits, and all subsystems continued to perform well.

At 12:48 p.m. EDT on August 5, after executing a roll and a pitch maneuver, the spacecraft fired its 100-pound-thrust rocket for 8 minutes and 28 seconds and decelerated by 643 meters per second into the gravitational captivity of the Moon. The initial orbital parameters were: apolune, 6,023 kilometers; perilune, 194.5 kilometers; inclination, 85.01°; period of orbit, 8 hours, 30 minutes. One and a half hours after orbit insertion, ground control commanded Lunar Orbiter V to scan the Goldstone test film, and the subsequent readout showed high-quality data. Following this, flight controllers prepared for the major photographic work of the mission.[41]

Photography commenced at 7:22 p.m. EDT on August 6. At this time the spacecraft took its first photograph of the Moon at a distance of about 6,000 kilometers from the lunar surface. The target was a previously unknown area of the far side. Then it executed a maneuver early on August 7

---

[41] Ibid. Lunar Orbiter V Post Launch Report #3, August 7, 1967.

that lowered the perilune to 100 kilometers while maintaining a 6,023-kilometer apolune. The spacecraft continued farside photography, exposing eighteen out of nineteen frames during the first part of the mission. The nineteenth was a "film set" frame, moved through the photo subsystem in an eight-hour interval to prevent film from setting and Bimat from drying out. While this was a planned item in the film's budget, the decision which program officials made early on August 7 changed the next scheduled "film set" frame significantly. They decided to use it to take a photograph of the Earth with the 610 mm high-resolution camera lens instead of passing it unexposed through the system.[42]

Site VA-9, as the Earth photograph was identified, had not been in the original plan. Program officals decided, however, that the position of <u>Lunar Orbiter V</u> relative to the Moon and the Earth and the Earth's position relative to the Sun afforded a very fine opportunity to take such a picture. The Langley program planning staff together with flight controllers implemented a plan to make an Earth photograph when the spacecraft neared apolune between orbits 7 and 8. Since the spacecraft's orbit geometry kept it in view of Earth at all times, the Moon would not appear in

---

[42] <u>Ibid.</u>, p. 2.

the photograph.[43]

Exactly seven hours twenty-three minutes elapsed between the exposure of the previous photograph of Site VA-8 and the moment when Lunar Orbiter V's camera made the historic picture of the nearly full Earth on August 8 at about 9:05 Greenwich Mean Time. Shutter speed was 1/100 second, but the Earth's high albedo caused some overexposure of the film. This was unavoidable. Later Langley Research Center photography specialists successfully applied image enhancement techniques, using magnetic tape video records of the readout of the photograph, to bring out details which would not have shown up in a negative reconstructed from the raw readout data. (Note that enhancement techniques did not involve any "doctoring" of photographic data in order to "show" something which was not there.)

Approximately 149° of arc of the Earth's surface appeared clearly in the photograph. It illustrated the possible synoptic weather observations that a satellite could conduct in cislunar space or that could be made from the Moon.[44]

---

[43] Lunar Orbiter V Photography, NASA CR-1094, prepared by the Boeing Company, June 1968, p. 140.

[44] Ibid., pp. 140-141. Picture and computer schematic on pp. 142-143.

Very early on August 9, EDT, <u>Lunar Orbiter V</u> executed a second orbital maneuver, which reduced its apolune from 6,023 kilometers to 1,500. The final orbital parameters were: apolune, 1,499.37 kilometers; perilune, 98.93 kilometers; inclination, 84.76°; period of orbit, 3 hours 11 minutes. All spacecraft subsystems continued to perform normally. The micrometeoroid detection experiment had recorded one hit, and the radiation level registered by the dosimeter at the film cassette remained constant at 1.0 rads, up from 0.75 rads.[45] In the following days the spacecraft continued to perform its mission as planned without experiencing any troubles. By August 14 it had completed 51 orbits and had exposed 107 of 212 film frames. Sixty frames had been read out, of which the picture of Earth showed remarkable detail from such a great distance.[46]

The photographic mission ended on August 18 when the spacecraft made its last photograph and ran out of Bimat at 11:20 p.m. EDT. In all it had successfully covered 5 Apollo sites, 36 science sites, 23 previously unphotographed areas on the lunar far side, and a view of the nearly fully illuminated Earth. The Apollo coverage included 5 sets of

---

[45] Post Launch MOR S-814-67-07, Lunar Orbiter V Post Launch Report #5, August 9, 1967.
[46] <u>Ibid</u>., #8, August 14, 1967.

convergent stereo photographs, each comprising two 4-frame sequences, and 4 westward-looking oblique views. <u>Lunar Orbiter V</u> had transmitted seventy-eight percent of the high-resolution photography to Earth at a rate of about 4 frames per orbit or 27 frames per day as of August 21, and ground control expected to conclude readout by August 26.[47]

The End of the Operational Phase

On September 2 Homer E. Newell, Associate Administrator for Space Science and Applications, certified that the fifth mission was an unqualified success according to prelaunch objectives. Deputy Administrator Robert C. Seamans, Jr., concurred on September 6. Both NASA officials also assessed the whole program as successful; five missions had been flown out of five planned.[48] Indeed the final Orbiter had capped an impressive effort by the Office of Space Science and Applications to bring man closer to stepping down upon the lunar soil and understanding where it was that he would be landing in the near future.

The status of the fifth Lunar Orbiter remained good following termination of readout early on the morning of

---

[47] Ibid., #10, August 21, 1967.

[48] NASA Mission Objectives for Lunar Orbiter E, signed by Edgar M. Cortright for Homer E. Newell, July 25 and September 2, 1967, and Robert C. Seamans, Jr., July 26 and September 6, 1967.

August 27. Lunar Orbiter II and III also continued to orbit the Moon and to provide extensive data on the lunar environment and its gravitational field. These three spacecraft served the Manned Space Flight Network as tracking targets for training personnel who would track Apollo.[49]

Lunar Orbiter II had sufficient attitude control gas to survive until early November. Ground control operators planned to impact it into the Apollo zone on the Moon's surface even though analysis of tracking data indicated that it could probably remain in orbit one or two years longer. Once the spacecraft lost its attitude control gas, however, it would become a derelict in orbit, beyond the control of ground operations. Program officials deemed it necessary, therefore, to crash the spacecraft while they could, to avoid any potential communications interference in future manned missions. They also planned to lower Lunar Orbiter III's apolune to make its orbit as circular as possible for further training for Apollo tracking. However, expiration of its gas would soon mean that it, too, would have to be crashed.

The fifth Orbiter had just begun its extended mission late in August. Its orbit would be changed on October 10 so

---

[49] Post Launch MOR S-418-67-07, Lunar Orbiter V Post Launch Report #11, September 7, 1967.

that it might better survive the umbral eclipse of October 18. (Program engineer Leon J. Kosofsky and mission operators changed the orbit so that the spacecraft would pass through the eclipse and solar occultation by the Moon at the same time.) Apollo network trackers would continue to track the spacecraft as long as possible to increase their experience in preparation for manned lunar missions.[50]

On September 11 the Lunar Orbiter Program Office issued a statement of the plans for terminating the life of the three remaining Orbiters. It stated briefly:

> The policy is to track the Orbiter spacecraft until the approach of loss of attitude control as indicated by the nitrogen pressure. While the spacecraft is still controllable, the engine will be fired so as to cause impact with the lunar surface. The impact will be made within the Apollo zone if feasible. At this time, it appears that Orbiter II will be impacted in early November, Orbiter III in mid October, and Orbiter V in mid summer 1968. Contact with Orbiter IV has been lost.[51]

Following the final acquisition of all Lunar Orbiter V photographic data, Lee R. Scherer issued a summary statement about the program's achievements. Among these he stressed that Lunar Orbiter II photography had led to the identification of the Ranger VIII impact point on the Moon. Orbiter

---

[50] Ibid.

[51] Lunar Orbiter Program Office, NASA, Termination of Active Lunar Orbiters: Present Plans for Terminating Active Lunar Orbiters II through V, Lunar Orbiter Item 29, September 11, 1967.

<u>III</u> photography had identified <u>Surveyor I</u> on the Moon's surface. The locations of the other Surveyors were also determined by using Orbiter photography. The fifth Orbiter had photographed major lunar features of scientific interest at a resolution 100 times better than Earth-based telescopes could achieve under ideal observation conditions. All Orbiters combined had photographed the entire lunar surface at a better resolution by at least an order of magnitude than Earth-based telescopes could attain and had surveyed the heavily cratered far side of the Moon. The spacecraft had provided valuable data contributing to the determination of the Moon's gravitational field. Finally, one of the program's most significant accomplishments had been to advance the Apollo Program in a way other than photographic site certification.

Five Orbiters had enabled the Manned Space Flight Network to train personnel in tracking and to check out equipment and computer programs for the manned lunar missions beginning with <u>Apollo 8</u> in December 1968 and including <u>Apollo 10</u> through <u>17</u>, of which all but <u>Apollo 10</u> and <u>13</u> landed on the Moon. (<u>Apollo 10</u> tested the complete spacecraft in lunar orbit and <u>Apollo 13</u> aborted its landing mission because an onboard oxygen tank exploded in cislunar space.) The Office of Manned Space Flight could not have obtained the needed tracking experience at a timely date if NASA had not flown the five Lunar Orbiter spacecraft.[52]

The chronology of the Lunar Orbiters concluded by the end of January 1968. On October 9, 1967, flight controllers commanded Lunar Orbiter III to impact on the Moon. On October 11 they commanded Lunar Orbiter II impact. They had lost communications with Lunar Orbiter IV on July 17, 1967, and assumed that its orbit had decayed sufficiently to permit it to crash onto the Moon late in October, but had no evidence confirming this. Lunar Orbiter V continued to fly its extended mission until, unexpectedly, it experienced an anomaly which threatened its orbit safety. A sudden loss of pressure in the nitrogen tank forced flight controllers to impact the spacecraft prematurely on the Moon to avoid losing it in orbit. They conducted this final maneuver on January 31, 1968, crashing Lunar Orbiter V near the equator on the Moon's western limb. The impact brought the operational phase of the Lunar Orbiter Program to a close.[53]

---

[52] Memorandum from SL/Assistant Director for Lunar Flight Programs (Lee R. Scherer) to SL/D. Pinkler, Subject: Lunar Orbiter Program Highlights, September 13, 1967, pp. 1-2.

[53] Information from Lunar Orbiter Program Office, NASA Headquarters; Lunar Orbiter Project Office, Langley Research Center; and Lunar Orbiter V Extended Mission Spacecraft Operations and Subsystem Performance, NASA CR-1142, prepared by the Boeing Company, August 1968, p. 121.

CHAPTER XI

CONCLUSIONS: LUNAR ORBITER'S CONTRIBUTION TO SPACE EXPLORATION

A Sixth Orbiter Mission?

Even before Lunar Orbiter V flew, the Office of Space Science and Applications was entertaining the prospect of flying a sixth Orbiter mission. Boeing had nearly enough parts to assemble another spacecraft at an initial cost of about $13 million. A gamma-ray experiment also existed which scientists desired to fly on a sixth Orbiter. Its inclusion would raise the cost of the mission by about $3 million. However, the necessity to relocate personnel on the Lunar Orbiter team to other jobs presented a major problem blocking another mission.[1]

Lunar Orbiter Program officials estimated that if the mission of Lunar Orbiter V failed, the program would have to fly a sixth Orbiter. However, refurbishment of a sixth spacecraft required such parts as two new solar panels. The Lunar Orbiter Program Office examined the needs and the lead times required for a sixth mission during May and June 1967. By the beginning of July, program management knew that OSSA soon had to make a commitment to another mission if it wanted

---

[1] Lunar Orbiter Program Office, NASA, Comments on Seamans Draft Memo (Undated), June 26, 1967. See also memorandum from SL/Manager, Lunar Orbiter Program, to SL/Acting Director, Lunar and Planetary Programs, Subject: Lunar Orbiter 6, April 6, 1967.

to avoid major shifts of personnel at Langley and Boeing following the photographic phase of Mission V. Known, too, was the simple fact that the longer NASA officials waited to approve the go-ahead for a new mission, the greater the costs and the more severely the management arrangements would impact on other NASA programs.[2]

On July 5 Scherer issued a statement summarizing the objectives of the fifth mission and the rationale behind a sixth Orbiter flight. He pointed out that the total cost of each of the first five missions amounted to $40 million apiece. The sixth mission would cost less than one third of this. Even if the fifth mission successfully achieved all planned objectives, a sixth mission could accomplish very valuable and different goals. Briefly it could 1) perform a total survey of the far side of the Moon at 60 to 80 meters resolution, 2) take a concentrated look at the best Apollo Program sites as determined through analysis of photographic data from the fifth mission, and 3) closely survey additional areas of high scientific interest. If Mission E failed,[3] a Mission F would be necessary, according to Scherer.[4]

---

[2] Ibid.

[3] NASA missions and spacecraft are denoted by capital letters (Mission E) during the prelaunch phase. After a successful launch, the mission and spacecraft are designated by numerals (Mission V).

[4] Lunar Orbiter Program Office, NASA, Action Item Summary, Action Item 31, Lunar Orbiter: Review and report the necessity for an additional Lunar Orbiter Mission, memo date June 16, 1967.

The Lunar Orbiter Project Office at Langley sent a memorandum to Scherer's Office on July 12 detailing the options open to OSSA for a sixth mission. The first option required a go-ahead decision by mid-July. The details were these: 1) that refurbishment and processing the spacecraft required four months and was the pacing item; 2) cost of launching Lunar Orbiter F late in November would amount to $12.75 million; 3) a launch by that time would retain the launch readiness capability of the previous launches; 4) this option provided the greatest retention of overall experience in the Lunar Orbiter team.[5] The second option was the same as the first except that it allowed for cancellation of preparations for a sixth flight early in September. At that time, data from Lunar Orbiter V would be available. If the mission was successful and the need for another mission was insufficiently justified, then the Lunar Orbiter Program could cancel the additional mission at a cost of about $4 million.[6]

The third option was the least manageable. It required that NASA postpone the July go-ahead but authorize funds to hold the team and the hardware in readiness until evaluation of the Lunar Orbiter V mission results. This option would

---

[5] Memorandum from Lunar Orbiter Project Office to NASA, Code SL, Attention: Capt. L. R. Scherer, Subject: Lunar Orbiter Project Recommendation for Implementing an Additional Mission, July 12, 1967.

[6] Ibid., p. 2.

extend the earliest possible launch date from late November 1967 to late January 1968 and raise the cost of a sixth mission to $16.5 million. It would also impact on the launch of OGO-E (Orbiting Geophysical Observatory satellite E) and would delay the Air Force takeover of Launch Complex 13 at Cape Kennedy. In view of these circumstances the Langley Lunar Orbiter Project Office recommended that only the first option be considered and that NASA Headquarters approve go-ahead before July 22, 1967.[7]

On July 14, 1967, Homer E. Newell sent NASA Deputy Administrator Robert C. Seamans, Jr., a summary of the alternatives for a sixth mission. He reiterated the three options which the Langley memorandum had specified and underlined Langley's position in support of a July go-ahead for a late November launch. He stressed to Seamans that a delayed decision would affect management problems, costs, and schedules in the Office of Space Science and Applications.[8]

Seamans weighed the need for a sixth mission and decided that NASA funds would better support other activities. On July 24, 1967, Scherer officially informed Langley that NASA Head-

---

[7] Ibid.

[8] Memorandum from S/Associate Administrator for Space Science and Applications to AD/Deputy Administrator, Subject: Considerations related to decision on a sixth Lunar Orbiter, July 14, 1967.

quarters had decided against a sixth Lunar Orbiter mission. However, he stated in his telegram to Floyd L. Thompson that a remote possibility for a reversal existed if the fifth mission failed. He requested Langley to proceed to phase out the program but to retain mission-peculiar test, launch, and flight operations equipment until it had completed the photo readout of Mission V. This retention did not apply to personnel, and Langley was to commence reassignment.[9]

Because <u>Lunar Orbiter V</u> succeeded beyond expectations in carrying out its mission objectives, its achievements proved that the cancellation of a sixth mission had been a prudent move. Moreover, the Apollo Program had virtually no need for the kind of data a sixth mission might have obtained; it would not have been decisive in mission planning. Indeed, at the Apollo Site Selection Board meeting on March 30, 1967, Apollo Program officials agreed that, "although further data from Lunar Orbiters D and E will be requested, the photography already received from Orbiters I, II, and III meets the minimal requirements of the Apollo Program for site survey for the first lunar landing."[10] They arrived at this conclusion

---

[9] Telegram, priority, unclassified, from Lee R. Scherer, Manager Lunar Orbiter Program, to Langley Research Center, Attention: Dr. F. L. Thompson, Mr. E. C. Draley, Mr. C. H. Nelson, July 24, 1967.

[10] Memorandum from MA/Apollo Program Director, Subject: Minutes of the Apollo Site Selection Board Meeting, March 30, 1967, p. 5.

by detailed screenings of Lunar Orbiter data using the following steps:

1. Construct Lunar Module landing ellipses and radar approach templets from photo support data.

2. Outline reject areas on medium resolution photographs.

3. Scan remaining area where high-resolution coverage is also available.

4. Select better ellipse locations with favorable radar approaches. Identify obstacles.

5. Select best ellipse based on landing and radar obstacles, count craters, and compute 'N' number from medium-resolution photos. For most favorable sites continue evaluation with high-resolution photography.

6. Evaluate ellipses on high-resolution photography and compute 'N' number.[11]

Apollo Mission Planning and Lunar Orbiter Data

The Apollo Program was the primary user of Lunar Orbiter data in the months following each Orbiter mission and in the period between the final mission and the first manned landing on the Moon in 1969. The story behind the Apollo site selection activities is beyond the scope of this history, but a brief summary of Lunar Orbiter's part in Apollo mission planning will demonstrate the role that

---

[11] *Ibid.*, Attachment--Steps in Lunar Orbiter Screening.

the Lunar Orbiter Program played in the Apollo Program as a result of cooperation between the Office of Space Science and Applications and the Office of Manned Space Flight.

The Apollo Site Selection Board (ASSB) had begun its work at its first meeting on March 16, 1966. No Lunar Orbiter or Surveyor spacecraft had yet flown, and, therefore, all discussion of site selection requirements had depended upon Ranger and Earth-based telescopic photography. Lunar Orbiter would soon change Apollo Program thinking about landing sites. At the first ASSB meeting the members identified a number of potential sites with the expectation that the sites finally chosen would be among them.[12]

By the following ASSB meeting Surveyor I had successfully landed on the Moon in Oceanus Procellarum, north of the crater Flamsteed. The first Lunar Orbiter mission, scheduled for early August, would attempt to photograph the Surveyor. Lunar Orbiter Program officials would adjust the positions of sites A-9 and A-10 to combine two blocks of photography for greater surface coverage of the area in which the unmanned spacecraft had touched down. In addition to this change in the first Lunar Orbiter mission, Norman Crabill and Thomas Young of the Lunar Orbiter Project Office, Langley, on June 1

---

[12] Memorandum from MA/Apollo Program Director, Subject: Minutes of Apollo Site Selection Board Meeting, March 16, 1966, document dated May 5, 1966.

presented the ASSB meeting recommendations for Lunar Orbiter Mission B. They believed that each Mission B site contained areas smooth enough to qualify as candidate Apollo sites. Finally the Apollo Program representatives, after reviewing the target sites for Lunar Orbiter Missions A and B, concluded that these sites would satisfy all known requirements for the Apollo missions if the surface of the Moon proved hospitable at each one.[13]

At the June 1 meeting Oran W. Nicks of OSSA asked Apollo Program people if they had any requirements for lunar landmarks which Orbiter could photograph. Owen E. Maynard of the Manned Spacecraft Center, who had presented the Apollo Site Selection Plan to the meeting, replied that the program had no plan at the time to use landmarks for updating orbits of the Apollo spacecraft. However, it would be desirable if such landmark sites could be located within a block of Orbiter photography containing a proposed Apollo landing site.[14]

By the December 15 ASSB meeting *Lunar Orbiter I* had obtained medium-resolution stereo photography of nine potential Apollo landing sites. *Lunar Orbiter II* had photographed thirteen potential sites in medium-resolution

---

[13] Minutes of the Joint Meeting of the Apollo Site Selection Board and the Surveyor/Orbiter Utilization Committee, June 1, 1966, document dated July 1, 1966, pp. 1-2.

[14] *Ibid.*, p. 3.

stereo and high-resolution monoscopic photography. Lawrence Rowan of the United States Geological Survey interpreted to those present the data of the lunar surface with respect to impact craters, volcanic fields, and mass wasting of the top layer of the Moon's soil. He made the following points in his talk:

1. Older mare areas such as those in <u>Lunar Orbiter II</u> photographs of Site II P-6 do not have the problem of crusts and lava tubes as young areas such as Site II P-2 most likely have.

2. <u>Surveyor I</u> photographs in Oceanus Procellarum exhibit more surface rocks than are found in Sinus Medii and Mare Tranquillitatis, suggesting that it might be younger and have a thin surface layer.

3. Slopes in older highland and smoothed mare craters, which show "patterned ground," may be unstable, with collapse or landslide dangers.[15]

Analysts for the Lunar Orbiter and Apollo Programs had chosen nine sites from <u>Lunar Orbiter I</u> photography and had applied Apollo site selection criteria in the effort to find Lunar Module landing areas. The December 15 ASSB meeting reviewed the results. Twenty-three areas proved large enough to contain a landing ellipse. These were undergoing further

---

[15] Minutes of Apollo Site Selection Board, December 15, 1966, document dated March 7, 1967. Site II P-6 is located in the southwestern area of Mare Tranquillitatis (approximately $23°$ east longitude, $2°$ north latitude). This site eventually became the <u>Apollo 11</u> landing site, Tranquility Base. Site II P-2 is located east of the crater Maskelyne and northeast of the crater Censorinus (approximately $33°$ east longitude, $3°$ north latitude).

study, and Apollo Program personnel evaluating them would make detailed crater counts of each during the next stage of selection. Following the preliminary analysis eight of the twenty-three areas merited special study.[16] The process of screening the Lunar Orbiter data is given in the diagram on the next page.

Landing site data determined from further analyses of Orbiter photography brought more confirmation that the Lunar Module design was correct and offered sufficient capability to land on the Moon. At a March 30, 1967, meeting of the ASSB, Donald C. Cheatham from the Manned Spacecraft Center pointed out that "the LM redesignation capability permits a change of touchdown point of 10,000 feet crosstrack at high gate (90 feet per second delta V, command at 30,000 feet down range). Visibility restrictions do not permit up-range redesignation. Preliminary examination of the Lunar Orbiter photography indicate that this capability will be sufficient for crater avoidance."[17] Already Lunar Orbiter had told Apollo mission planners much about the areas where they could and could not send a Lunar Module.

---

[16] Ibid., Attachment G, Preliminary Landing Site Analysis of Orbiter I, p. 2.

[17] Minutes of the Apollo Site Selection Board Meeting, March 30, 1967, document dated June 26, 1967, p. 1.

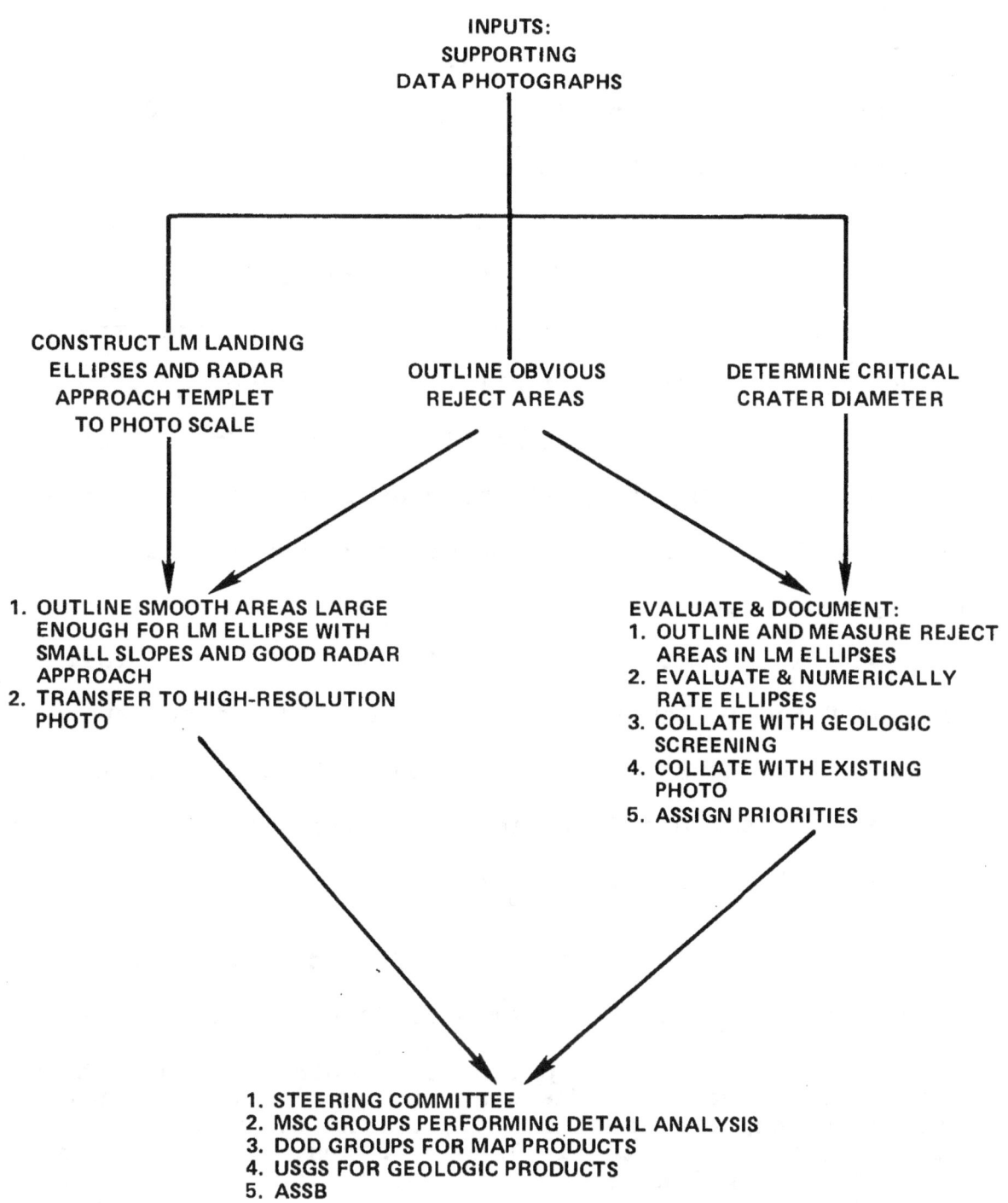

Source: Minutes of the Apollo Site Selection Board Meeting, December 15, 1966, Attachment G, p. 3.

PRELIMINARY SCREENING

Finally, the December 15, 1967, meeting of the ASSB at Houston had the photographic data of all five Lunar Orbiters upon which to base its judgments. The major criteria for selection of the landing sites subsequently depended upon performance constraints of the Apollo spacecraft, particularly the Lunar Module.[18] Lunar Orbiter had provided the photographic data which the Apollo Program had originally requested. Surveyor data continued to come in from three landed spacecraft in the Apollo zone of interest. Two more Surveyors would land in different areas of the Moon before that program concluded operations. Beyond this, Lunar Orbiter photography did not constitute a major basis for the final selection of Apollo landing sites. Selection had to depend upon performance constraints of the Lunar Module. At this point Lunar Orbiter had fulfilled its primary mission for the Apollo Manned Lunar Landing Program.

A year later, after the first Apollo mission to orbit the Moon, Apollo 8 Astronaut James A. Lovell, Jr., reported:

> . . . the Lunar Orbiter photographs which we had on board were quite adequate. There was no problem at all in determining objects, particularly on the near side of the moon. There are suitable landing sites. They are very easily distinguished. We could pick them up. We could work our way in. . . . The Lunar Orbiter photos again were helpful . . . to check the craters on the back side.[19]

---

[18] Minutes of the Apollo Site Selection Board Meeting of December 15, 1967, document dated January 29, 1968.

[19] Manned Spacecraft Center, Apollo 8 Technical Debriefing, January 2, 1969, p. 34.

Results of Non-photographic Lunar Orbiter Experiments

The micrometeoroid experiments flown on all five Lunar Orbiter spacecraft provided measurements in the near-lunar environment of the rate of penetration by meteoroids of 0.025-millimeter beryllium copper detectors. Each spacecraft carried twenty detectors, totaling an effective exposed area of 0.186 square meter. The spacecraft flew both equatorial and polar orbits at altitudes ranging from 30 to 6,200 kilometers and collected data on micrometeoroid impacts for a period of seventeen months.[20]

A primary goal of the experiment was to obtain data for the purpose of comparing the meteoroid hazard near the Moon with that near the Earth. These data would aid the Apollo Program in the determination of the amount of protection necessary for the spacesuits, instruments, and spacecraft. Moreover, they would refine the estimates of the hazard in near-lunar environment which scientists had made and which ranged from somewhat less to greater by several orders of magnitude than the hazard near the Earth. A major uncertainty was the contribution of secondary meteoroids created by the impacts of primary meteoroids on the Moon.

Before the Lunar Orbiter missions, only the Soviet lunar satellite _Luna X_ had measured meteoroid flux near the Moon. The results of its experiment showed that the average

---

[20] Charles A. Gurtler and Gary W. Grew, "Meteoroid Hazard near Moon," _Science_, Vol. 161 (August 2, 1968), p. 462.

rate of flux of micrometeoroids exceeded the average for interplanetary space by about two orders of magnitude.[21]

To arrive at the average rate of flux for the five Lunar Orbiter micrometeoroid experiments, the primary investigators (Gurtler, Kinnard, and Grew) divided the total number of recorded punctures by the total time-area product. The five Orbiters recorded 22 punctures during a time-area exposure of 139.0 square meters per day. These figures gave an average rate of 0.16 puncture per square meter per day ($m^2 \times day^{-2}$) in the near-lunar environment, or about one half the average rate of flux recorded by the Earth-orbiting satellites Explorer XVI and Explorer XXIII:[22]

Spacecraft	Punctures	Exposure ($m^2 \times day$)	Punctures ($m^{-2} \times day^{-1}$)
Lunar Orbiter I through V	22	139.0	0.16
Explorer XVI	44	132.9	.33
Explorer XXIII	50	139.9	.36

The investigators found by analysis of the 22 punctures of the micrometeoroid detectors, in relation to spacecraft positions at time of impact, that there was a preponderance of punctures on the side of the spacecraft facing forward in

---

[21] T. N. Nazarova, A. K. Rybakov, C. D. Komissarov, "Investigation of solid interplanetary matter in the vicinity of the Moon," paper before 10th COSPAR meeting, London, July 1967.

[22] Gurtler and Grew, "Meteoroid Hazard near Moon," p. 463.

the orbital direction of movement around the Sun. This preponderance agreed with Earth-based radar observations cited by G. S. Hawkins [23] and indicated that the influx of meteoroids on the side of Earth facing forward in orbit around the Sun was several times greater than influx on the opposite side.[24]

Preliminary estimates of the flux of secondary meteoroids near the Moon indicated that flux was greater nearer the lunar surface and dropped off sharply with increase in altitude.[25] Further study of the Lunar Orbiter data indicated no statistically significant variation of hazard with altitude.

Gurtler and Grew concluded, in the summary of their analysis of the micrometeoroid experiment data, that the penetration rates in the near-lunar environment as well as near the Earth should be accepted as being only tentative since the number of recorded penetrations was statistically small and the meteoroid flux near the Earth's orbit might vary from one measurement period to another. However, the data did indicate that the penetration hazard for 0.025 millimeter of beryllium copper was no greater near the Moon than near the Earth. Nor

---

[23] G. S. Hawkins, *Monthly Notices of the Royal Astronomical Society*, Vol. 116, No. 1 (1956), p. 92

[24] Gurtler and Grew, "Meteoroid Hazard near Moon," p. 463.

[25] D. E. Gault, E. M. Shoemaker, and H. J. Moore, *Fragments Ejected from Lunar Surface by Meteoroid Impact Analyzed on Basis of Studies of Hypervelocity Impact in Rock and Sand*, NASA Technical Note TN-D-1767, 1963.

was there any substantial evidence that the hazard in the near-lunar environment increased as a result of secondary meteoroid impacts caused by primary impacts on the Moon.[26]

The data obtained from the radiation experiments on board the five Lunar Orbiter spacecraft had significant implications for the Apollo Program. What would be the approximate doses of radiation experienced by astronauts in space suits? In the Lunar Module? In the Apollo Command Module? To obtain an answer, the primary investigator, Dr. Trutz Foelsche, analyzed the data recorded by the two cesium iodide (CsI) detectors in each of the five Orbiters. One of the two was shielded by 0.2 gram of aluminum per square centimeter, the other by 2.0 grams aluminum per square centimeter. Because of the higher absorption of protons and alpha-particles per gram per square centimeter in soft tissue or water, the doses recorded by the Lunar Orbiter dosimeters had to be multiplied by two. The analysis showed that all events recorded were of significance to a man in space only where shielding was light, specifically in a space suit or in the Lunar Module.[27]

The following table shows the skin doses that would be incurred in a space suit with shielding of 0.17 gram per square centimeter in the presence of three solar particle events.[28]

---

[26]Gurtler and Grew, "Meteoroid Hazard near Moon," p. 464.
[27]Foelsche, "Radiation Measurements in LO I-V," p. 7.
[28]Ibid.

Event Date	Radiation Dosage
September 2, 1966	270 rads in H_2O
January 28, 1967	106 rads in H_2O (24 rads behind 2 grams/cm^2 shielding)
May 24/28, 1967	130 rads in H_2O (<u>Lunar Orbiter IV</u> in high orbit)

Foelsche noted, however, that the skin doses approached or even surpassed the suggested maximum permissable skin dose (MPD) for astronauts for short-term exposure even for the moderate rates above. See the table below.[29]

Types of Mission	Suggested MPDs for Astronauts		
	Eyes (rad)	Blood-Forming Organs (rad)	Skin (rad)
Short Term (up to two weeks)	27	52	233
Long Term (several months)	250	150	500

In summary, the Lunar Orbiter radiation experiments contributed to four areas of scientific interest in addition

---

[29] <u>Ibid</u>.

to monitoring the doses on the camera film. First, they allowed estimates to be made of the skin dose rates behind 2 grams per square centimeter of shielding for astronauts passing through the Van Allen Belt. The estimates made from these data were based on an assumption of five passes through the belt in a one-year period. Second, the experiments contributed to information about the Moon's core. The weakness or absence of an intrinisic magnetic field of the Moon, which Explorer XXXV confirmed, indicated that the Moon has no extended liquid conducting core like that scientists accept for the Earth.

Third, by comparing data of Pioneer V and VI (spacecraft that lagged behind or were ahead of the Earth while in orbit around the Sun) with Lunar Orbiter data, preliminary conclusions could be drawn concerning the spatial and lateral extensions and the intensities of solar particle flux during the 1966 and 1967 events. Finally, the experiments measured, by simulation, high skin doses in a light space suit near or on the Moon for the moderate size solar particle events of the August 1966 to August 1967 time span. From these data the inference could be made that in rare cases of large event groups, such as those of 1959 and 1960, the Apollo astronauts might experience skin doses greater than 1,800 to 5,000 rads in one week, if no precautions were taken.[30]

The radiation experiments produced data which confirmed that the design of the hardware that Apollo astronauts would use on their lunar missions beginning in 1969 would protect them from average and greater than average short-term exposure to solar particle events.

A Meaning for the Lunar Orbiter Achievements

Doubtless much more can be said about the Lunar Orbiter Program and its relationship to Apollo. However, this must be the task of future historians of space exploration. It now remains for this author to draw his conclusions about the Lunar Orbiter Program. These are certainly preliminary, and any error must be attributed to the author.

The Lunar Orbiter Program, like the Apollo Program, had unfolded in a politically charged atmosphere. The national commitment to land Americans on the Moon within the decade of the sixties imposed certain directions and a sense of urgency on the course which the National Aeronautics and Space Administration took in both programs. It also placed certain limitations on unmanned exploration of the Moon. First, the Apollo Program provided Lunar Orbiter with its raison d'être. This meant that the Office of Space Science

---

[30] Ibid., pp. 7-8.

and Applications undertook an engineering feat in 1963 whose most immediate applications would directly support the objectives of the Apollo Program, to design and build a system and a mission that could take men to the Moon and return them safely to Earth. Lunar Orbiter contributed significantly to Apollo mission design (the hardware been designed and built before the Lunar Orbiter mission operations began). In this it supplemented the pioneering work of Ranger and Surveyor.

      The American commitment for a manned lunar landing and the needs of Apollo eclipsed unmanned scientific exploration of the Moon during the sixties. The Office of Space Science and Applications thus also stood in the shadow of the Office of Manned Space Flight in lunar exploration. On the other hand, OMSF owed OSSA a debt of gratitude for the ground-breaking, precursory work that Ranger, Surveyor, and Lunar Orbiter did. Moreover, the highly successful Lunar Orbiter Program proved the role that unmanned, long-life orbiters could play in future space exploration. It is no coincidence that Langley Research Center, which directed the Lunar Orbiter Program, was in 1976 carrying out the operational phase of the Viking Mars program, with two Viking spacecraft on their way to orbit and land on Mars. Jet Propulsion Laboratory, the other major unit carrying out lunar and planetary exploration programs (Ranger, Surveyor, Mariner), also was playing a key role in Viking.

American exploration of the Moon obtained space-proved systems to conduct specific observations and to gather precise data on the lunar environment, with or without men. But altering national priorities, government belt-tightening, and reduced NASA budgets foreclosed lunar exploration after the *Apollo 17* landing in 1972, at least for this decade.

The once ambitious unmanned lunar exploration program, Surveyor Orbiter, which would have carried a wide variety of scientific instruments and experiments to the Moon's environment much as the Soviet Luna and Zond spacecraft have,[31] has not been attempted again. Perhaps it was too ambitious for its time; and the road taken to land men on the Moon proved politically more reassuring.[32] Certainly the five out of five successful missions of Lunar Orbiter and the desire to fly a sixth mission substantiated the philosophy within NASA that unmanned lunar probes served best when their objectives were simple, limited, and mutually supportive of each other and of manned exploration.

Had the Office of Space Science and Applications directed the five missions of Lunar Orbiter to conduct scientific

---

[31] See Record of Unmanned Lunar Exploration Probes, Appendix C.

[32] John M. Logsdon gives a detailed and documented account of the decision-making process behind initiation of a manned lunar landing program in his book *We Should Go to the Moon* (Cambridge: Massachusetts Institute of Technology Press, 1970).

investigations of the Moon, independently of Apollo, then most likely the missions would have been different. Mission IV might have been the first to fly. A total survey of the Moon would have allowed scientists to select the most interesting sites for closer, more detailed photographic investigations. Surveyor spacecraft might have landed elsewhere than they did, because of Lunar Orbiter data; and even Apollo might have flown significantly different missions. This, however, did not happen.

If Lunar Orbiter had been totally independent of any manned exploration, much as the Mariner Mars spacecraft have been, then perhaps only part of the missions would have flown photographic payloads. Numerous experiments to analyze the Moon's environment existed or could have been designed to fly on an Orbiter, as they were flown on *Explorer XXXV*. Yet Lunar Orbiter could not have satisfied the poli-commitments the United States had made as a result of the early Soviet thrust into space. In fact, Lunar Orbiter was inseparably bound to the goals of the American manned lunar exploration effort.

The bond between Lunar Orbiter and Apollo fostered cooperation between the Office of Space Science and Applications and the Office of Manned Space Flight, which otherwise might have developed more slowly and less affirmatively. This cooperation brought about a higher level of integrated activities among NASA centers far sooner than might have occurred under different circumstances. The problems encountered in the Ranger and Surveyor Programs early in the sixties forced NASA Headquarters to search for other means of accomplishing the tasks of space exploration, leading it to delegate to the Langley Research Center a new area of responsibilities beyond its traditional role in research and development. In turn this move has broadened the agency's base for accomplishing ever more complex and sophisticated objectives in American space exploration.

It would be unjust, however, to claim that without Lunar Orbiter photography, Apollo could never have flown so early or that America could not have landed on the Moon in 1969. Lunar Orbiter greatly illuminated Apollo's way, but it is highly conceivable that the Apollo Program could have flown one or more manned orbital photographic missions before planning a landing. No Orbiter data went into the design of the Apollo spacecraft system; and, indeed, the missions of <u>Apollo 8</u> and <u>10</u> demonstrated the orbital capabilities of the spacecraft. The main objective of these two missions was testing the systems and the mission design short of actual landing on the Moon. The photography by the astronauts on these missions was concentrated on landing sites. The Lunar Orbiter photography covered almost the entire Moon and captured scenes of the lunar landscape under predetermined lighting conditions and at altitudes that allowed Lunar Orbiter Program officials to obtain precise information about the landing sites, which the Apollo Program had requested. Moreover, it obtained these data at a time when they proved most useful to Apollo mission design.

Thus Lunar Orbiter saved Apollo time. It also saved money; the cost of one Apollo manned mission to the Moon was far higher than the total cost of the whole Lunar Orbiter Program. Without Lunar Orbiter, NASA might have had to fly one or more manned orbital missions around the Moon to photograph potential landing sites before an actual manned landing mission. Lunar Orbiter also gave Apollo flight operations personnel experience in tracking five spacecraft in orbit around the Moon. It provided valuable data on the lunar gravitational environment and its effects on orbiting spacecraft. It aided the Surveyor Program in selecting landing sites and then it photographed the landed Surveyors. Lunar Orbiter V photography of the crater Tycho and its vicinity proved instrumental in the decision to land Surveyor VII north of Tycho in an area of high scientific interest but with topography greatly reducing the chances of a soft landing. Surveyor VII landed successfully and provided valuable data on an area of the Moon where astronauts did not land. The teamwork of the Lunar Orbiter V and Surveyor VII missions demonstrated the value of unmanned lunar exploration.

The successful achievements of Lunar Orbiter and Surveyor also had far-reaching implications for planetary exploration. The former director of the OSSA Office of Lunar and Planetary Programs, Oran W. Nicks (later Deputy Director of the Langley Research Center), outlined some of these implications in an address to the American Institute of Aeronautics and Astronautics on December 5, 1968. He stated that experience gained in the initial stages of unmanned lunar exploration would have direct applications in the exploration of the planet Mars in the seventies.[33]

Exploration of Mars at close range began in 1965 with the fly-by of **Mariner IV**. It provided man his first detailed glimpse of the Martian surface; surprisingly its pictures revealed many craters, showing apparent similarities to the Moon. In July and August 1969, **Mariner VI** and **Mariner VII** brought even closer views of the red planet when they flew by, taking pictures and measurements of the atmosphere and surface temperatures. **Mariner IX** went into orbit of Mars in November 1971 and in one year of observations changed scientists' views of the planet's weather and possible evolution. These spacecraft have opened many more areas of questioning than they have answered and, as a result, the Viking Program would search for evidence of life

---

[33] Oran W. Nicks, "Applying Surveyor and Lunar Orbiter Techniques to Mars," address before the American Institute of Aeronautics and Astronautics, Washington, D.C., December 5, 1968, pp. 10-11.

on Mars during 1976-1977. Although the weight and payloads of the Viking Mars probes were to be substantially different from those of Lunar Orbiter, the spacecraft would profit from the Orbiter experience. The Viking Program at Langley and at JPL could use the knowledge gained from both Lunar Orbiter and Surveyor, although its goals required much more complex hardware and missions.

The Viking Program's relationship to Lunar Orbiter demonstrates how the Office of Space Science and Applications successfully built on the cumulative knowledge gained in its programs in the previous fifteen years. Among other achievements, this work proved the Orbiter concept and the feasibility of landing an unmanned spacecraft on another celestial body. Viking could draw on an an increasing treasury of proved concepts in furthering the unmanned exploration of the solar system. It also would add to that treasury. Nicks summed up the meaning of this work in his address to the American Institute of Aeronautics and Astronautics in December 1968:

> Burning questions of immediate concern to you and me will be addressed by use of our new tools: "Is there life elsewhere? Has life existed on nearby planets and disappeared for any reason? Can nearby planets be made suitable for life?"
>
> Together, orbiters and landers form a powerful team for the study of Mars and for seeking answers to these questions. Together, they will continue to extend our capabilities in what is probably the most challenging, open-ended arena for expansion of science and technology in the decade ahead.[34]

---

[34] Ibid., p. 12.

Men have now landed six times on the Moon and have returned with samples of its surface and subsurface materials. It still remains a mysterious body, its surface barely scratched; exploration of it has only begun. Mars, Venus, Mercury, and Jupiter have been studied by space probes and the other planets beckon men to pursue the quest for an answer to the origins of the Earth, the solar system, and, eventually, the universe.

Two Lunar Orbiter photographs had especially far-reaching implications for the Earth's population. The first was the Earth-Moon picture made in August 1966 by Lunar Orbiter I. Nearly half of the Earth was shown, as well as a substantial portion of the Moon's cratered surface.[35] The second was the Lunar Orbiter V picture of the nearly full Earth, taken in August 1967 while the spacecraft was at apolune in its nearly polar orbit of the Moon.[36] Both were unscheduled pictures, requiring extra planning to execute. Their success proved the versatility of the Lunar Orbiter spacecraft and the skill of the flight operations personnel, bearing witness to America's technological ingenuity and imagination.

---

[35] A detailed description of the Lunar Orbiter I Earth-Moon photography is given in Lunar Orbiter I--Photography, NASA CR-847, August 1967, pp. 64-71.

[36] For details of the Lunar Orbiter V Earth photograph, refer to Lunar Orbiter V--Photography, NASA CR-1094, June 1968, pp. 140-141.

Yet both photographs were unrelated to the Apollo manned lunar landing missions. They came two years before the first landing. Although not the first pictures of the Earth from space, they were the first to show Earth at the distance of its nearest neighbor.

To the historian they have perhaps a different meaning than to the scientist. Men, it seems, have always been on one quest or another, using the Moon, the Sun, the planets and the stars in varying ways to explain their existence and their destiny. Half a millenium ago Europeans believed the Earth to be flat and the center of the universe. Then slowly men such as Copernicus, Kepler, Galileo, and Newton altered the thinking about the universe. The old Aristotelian-Ptolemaic concepts of physics and astronomy that had, in part, shaped medievial man's thinking about his existence, dissolved in the new body of increasing empirical data on man's natural environment. Yet only yesterday were men able to see how finitely microscopic their home in space is.

Man's technology has enabled him to escape the Earth, land on the Moon, and return. It also has silently, visually warned him that his only home, for the present, is the blue-brown-white gem around which the cratered, desolate Moon revolves.

CHAPTER XII

LUNAR ORBITER PHOTOGRAPHY

The author selected the following pictures and their captions with the advice of Leon J. Kosofsky, former Lunar Orbiter program engineer, and Farouk El-Baz, formerly with Bellcomm, Inc., and now with the Smithsonian Institution. The selection offers a survey of the program's different phases. It does not constitute a scientific analysis of the Moon, but merely samples Lunar Orbiter photographic achievements.

For more detailed, analytical sources, the reader may refer to Leon J. Kosofsky and Farouk El-Baz, The Moon as Viewed by Lunar Orbiter, NASA SP-200, 1970; and J. Kendrick Hughes and David E. Bowker, Lunar Orbiter Photographic Atlas of the Moon, NASA SP-206, 1971.

The photographs of the Moon reproduced in this history do not represent the ultimate quality in Lunar Orbiter photography. They have been made from negatives of an unknown generation and therefore their actual resolution is uncertain; Lunar Orbiter photos reconstituted from original data had known resolutions. NASA has enhanced Orbiter photography for applications at Langley Research Center and in cooperation with the United States Geological Survey, U.S. Air Force Aeronautical Chart and Information Center, and Army Map Service.

Responsibility for any errors in the brief descriptions accompanying these photos must rest solely with the author.

A. THE SPACECRAFT

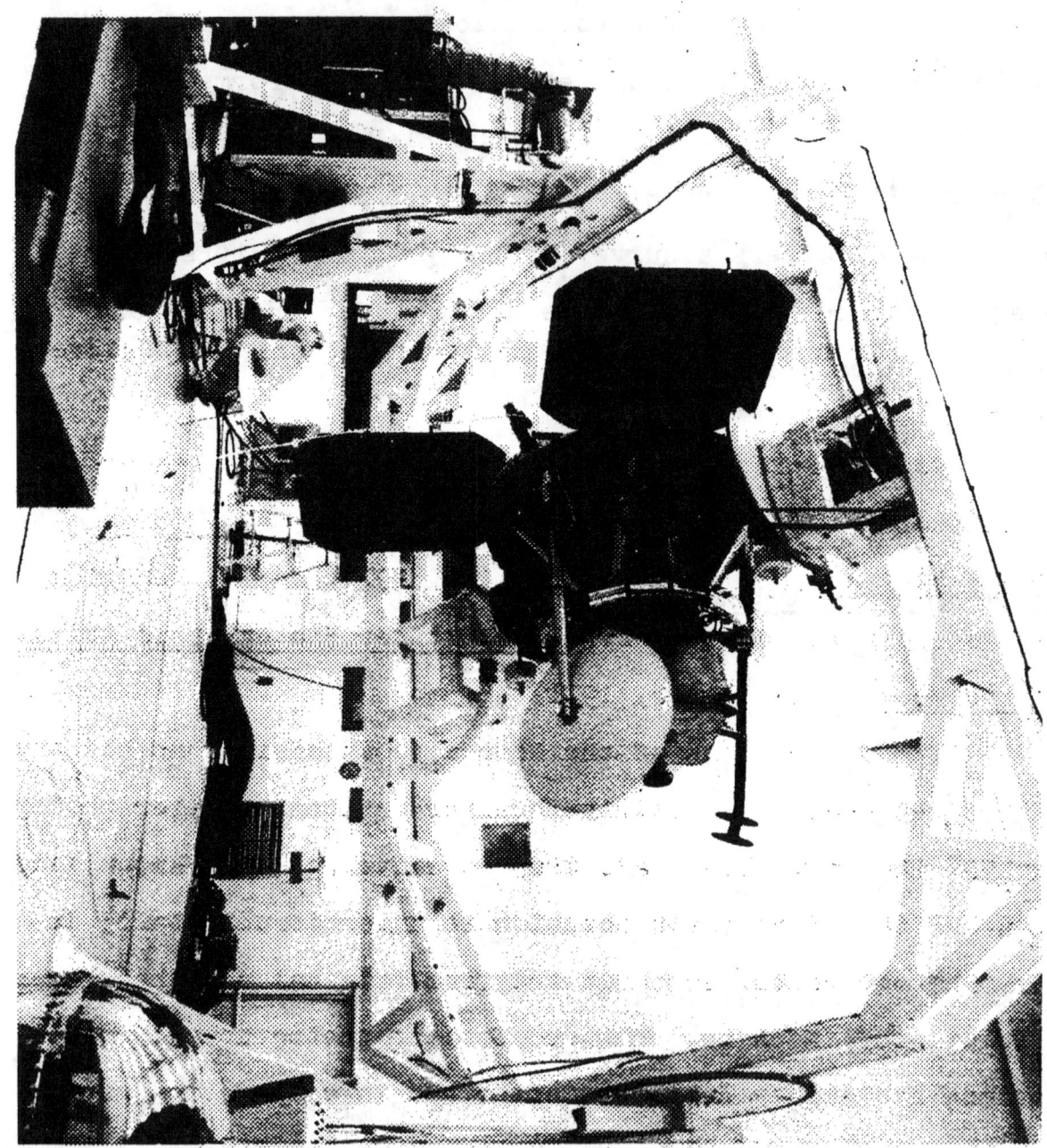

A Lunar Orbiter spacecraft in testing in NASA Hangar S Clean Room at Kennedy Space Center. The spacecraft was mounted on a three-axis test stand with its solar panels deployed. The one-meter-diameter high-gain dish antenna extended from the side of the Orbiter.

The spacecraft's main equipment deck and fuel tank deck held vital components. The back of the photographic system casing ("Bathtub") shows below the fuel tanks, and portions of the four solar panels that supplied power to the systems can be seen stretching from beneath the spacecraft.

The photographic system of <u>Lunar Orbiter V</u> undergoing tests at Cape Kennedy. Technological capability to compress all necessary components into an eggshell container with a total weight of less than 70 kilograms made the mission possible. The camera had two lenses: a wide-angle, medium-resolution 80 mm Xenotar Schneider-Kreuznach manufactured in West Germany and a 610 mm high-resolution telephoto Panoramic manufactured by Pacific Optical Company. Both were adjustable to the same exposure times of 1/25, 1/50, and 1/100 second. The Kodak special high-definition aerial film, Type SO-243, had a slow exposure index of ASA 1.6. It was extremely fine-grain film, requiring low shutter speeds, but was also less susceptible to radiation fogging. The lenses were protected by a quartz window and a metal door.

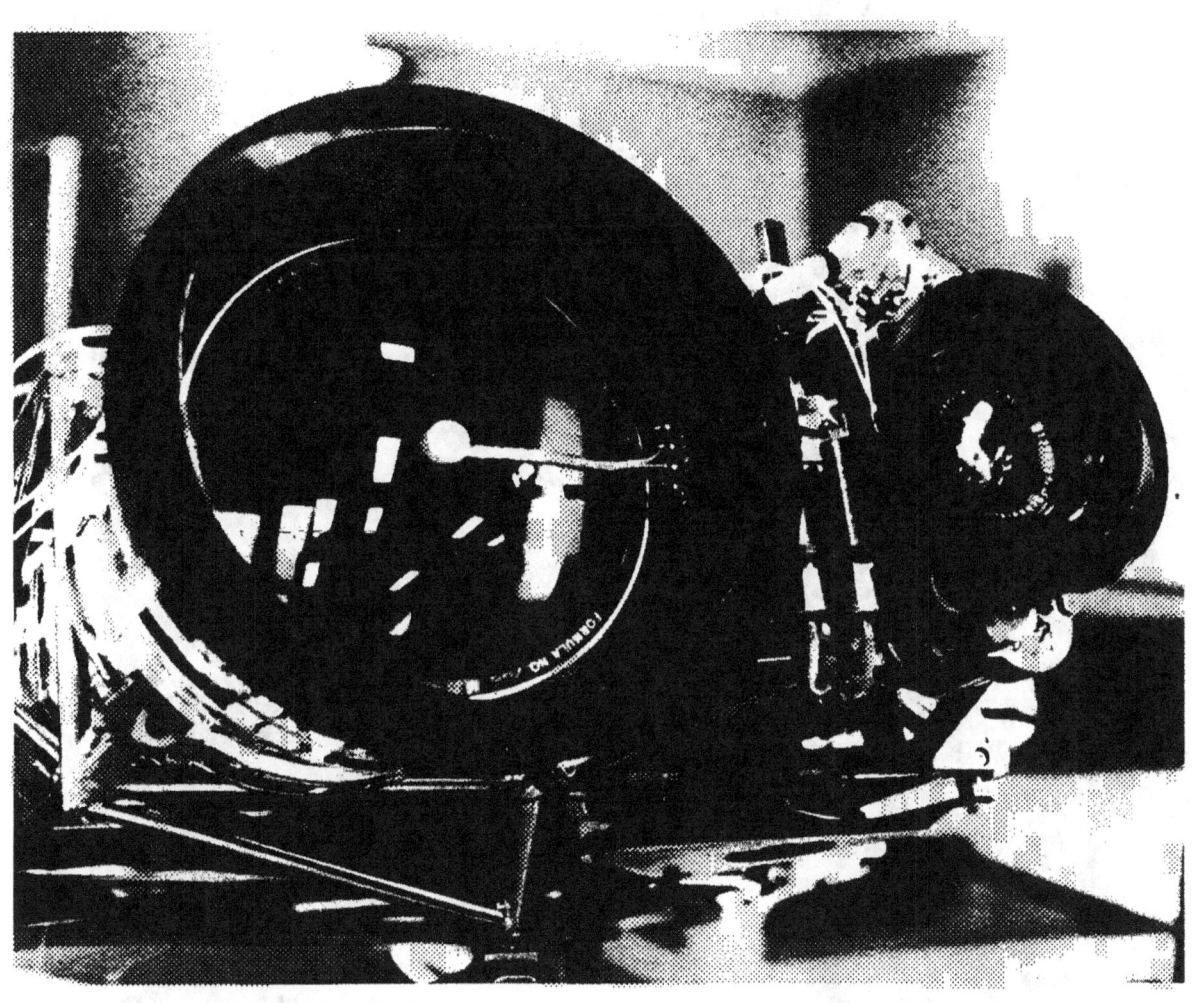

Close-up of the Eastman Kodak photographic system. The 610 mm F 5.6 high-resolution lens (left) and the 80 mm F 2.8 medium-resolution lens (right) gave the Lunar Orbiter a dual-imaging capability — the ability to take two kinds of pictures simultaneously on the same film.

The processor of the photographic system included three drums. The drum at the upper left held the Kodak Bimat web (processing film). The Bimat, covered with a gelatin layer saturated with a photographic processing solution, was laminated with the exposed camera film on the small drum in the center. In 3.5 minutes it developed and fixed the film. Then it separated from the film and wound onto the spoked take-up reel to the right of the small drum. The camera film passed over the large drying drum at the bottom, where it dried in 11.5 minutes at 35° C before moving to readout.

B. MISSION OPERATIONS

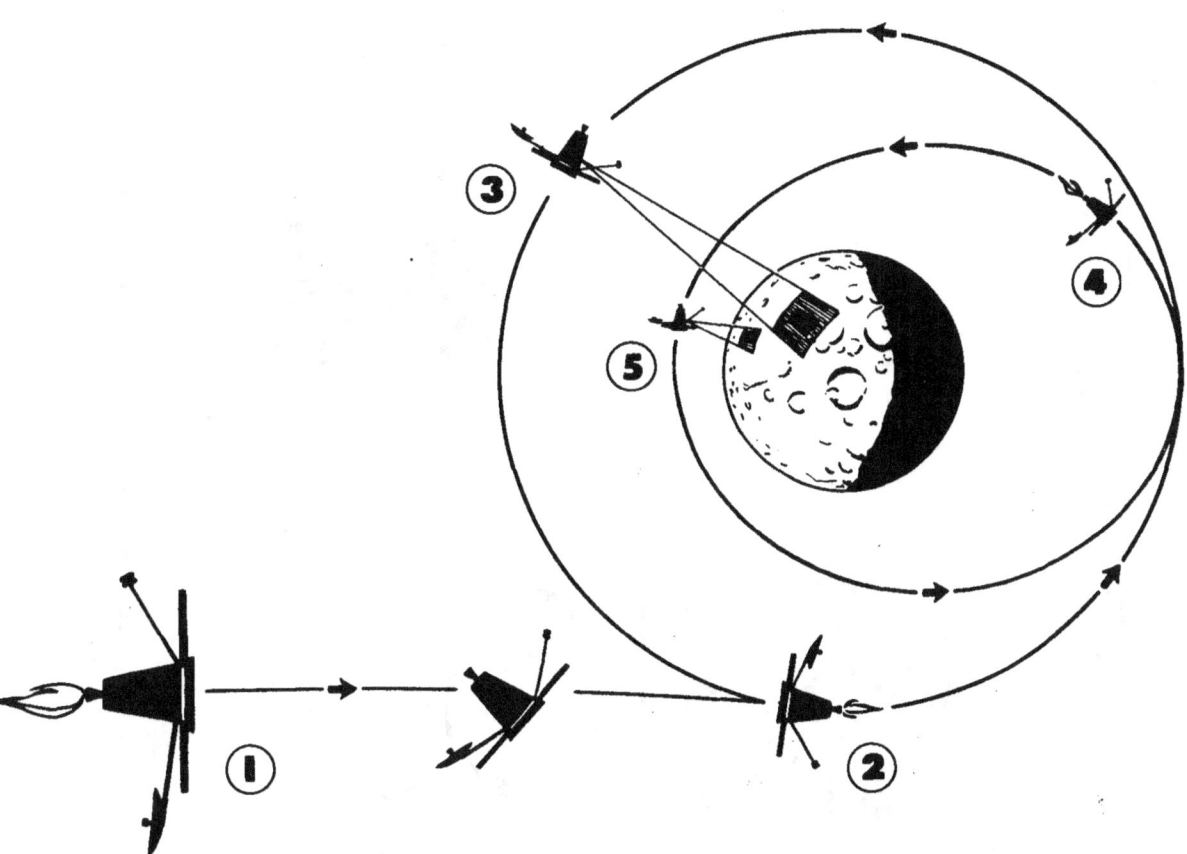

The drawing outlines essential steps in a Lunar Orbiter mission following launch and transit to the Moon's vicinity. In step 1 the spacecraft fired its velocity control rocket to make a course correction. In step 2 the rocket fired again to deboost the spacecraft into its initial orbit of the Moon. Here its orbit was adjusted, and the first pictures were made (3) before the Orbiter changed orbital parameters (4) to assume an elliptical orbit that brought it closer to the lunar surface for further photographic coverage (5).

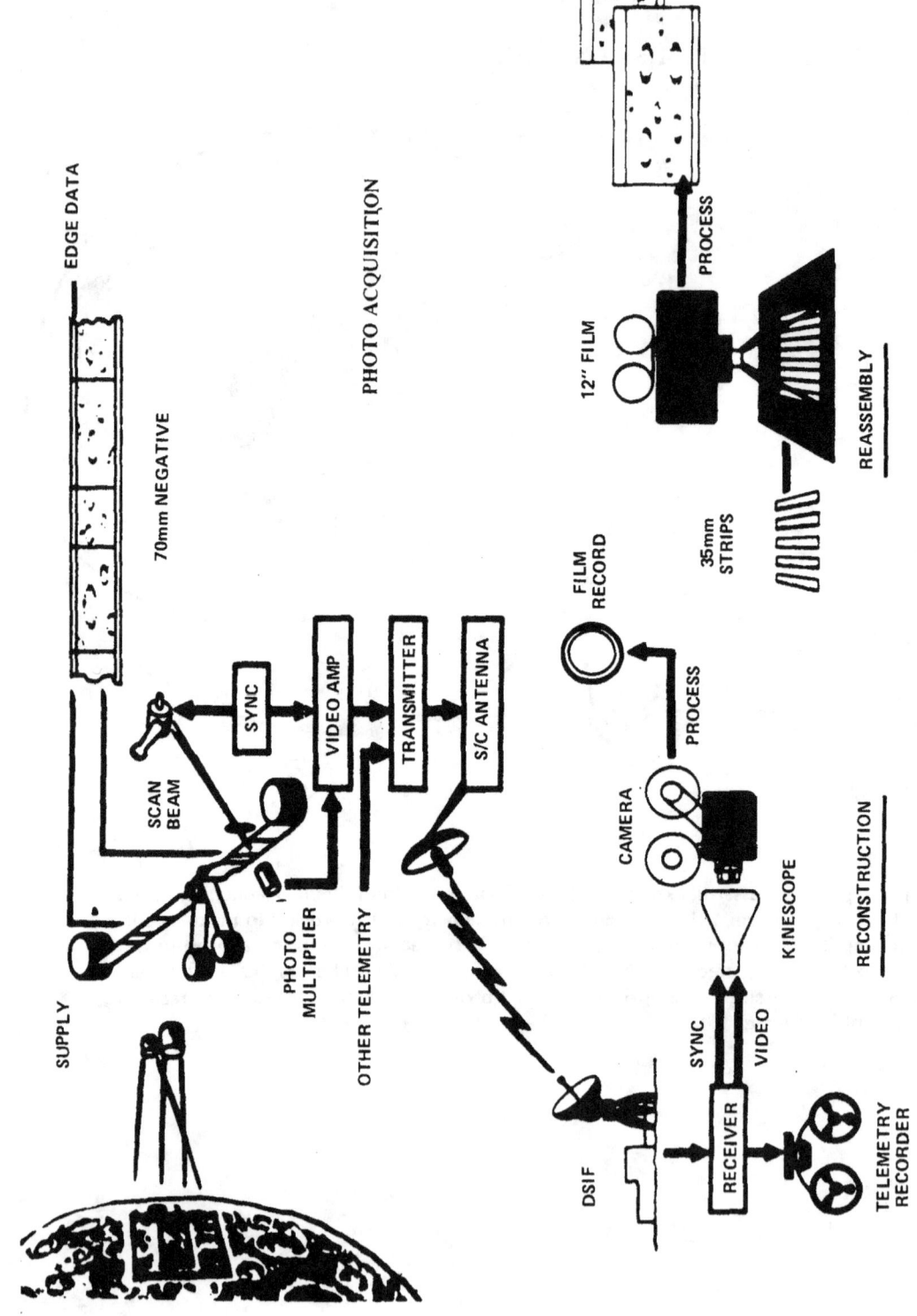

Steps in the acquisition of photographic data by Lunar Orbiter included transmission to Earth, readout, reconstruction, and reassembly for evaluation.

C. APOLLO SITE SURVEY

PRIMARY SITE 2 – GOOD

WORTHY OF FURTHER ANALYSIS

PRIMARY SITE 4 – POOR

REJECTED

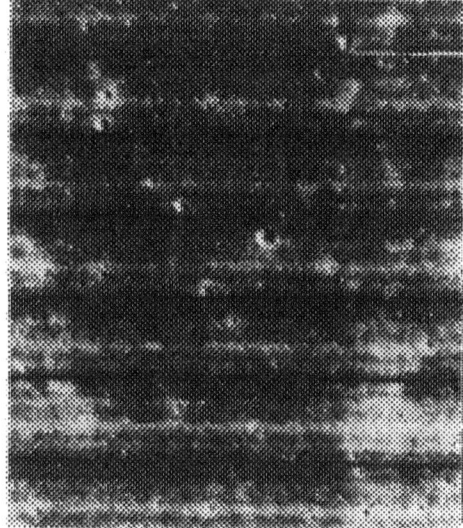
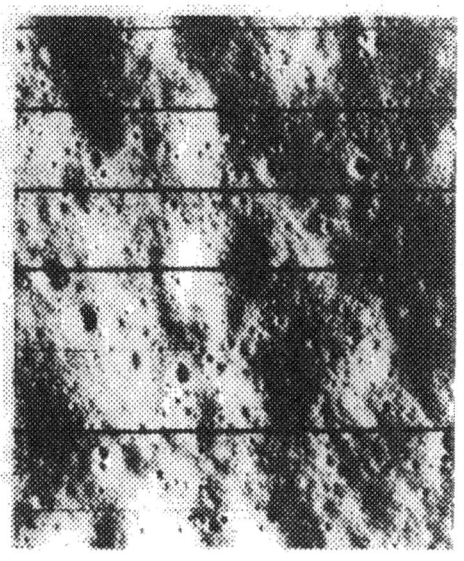

LUNAR ORBITER II APOLLO SITE SEARCH

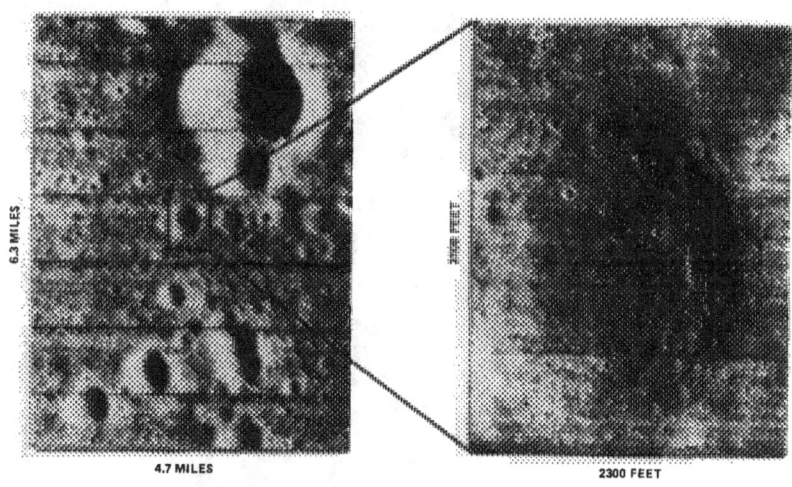

6.3 MILES

4.7 MILES

2300 FEET

SEA OF TRANQUILITY

ORBITER II RESULTS
SURVEY OF POTENTIAL APOLLO LANDING SITES

NASA S67-1997
2-24-67

Lunar Orbiter II photographed potential Apollo landing sites.

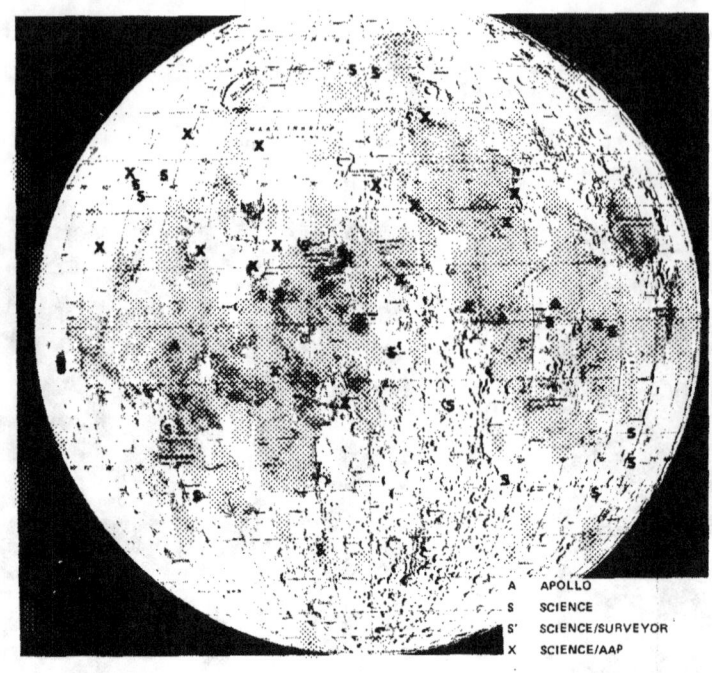

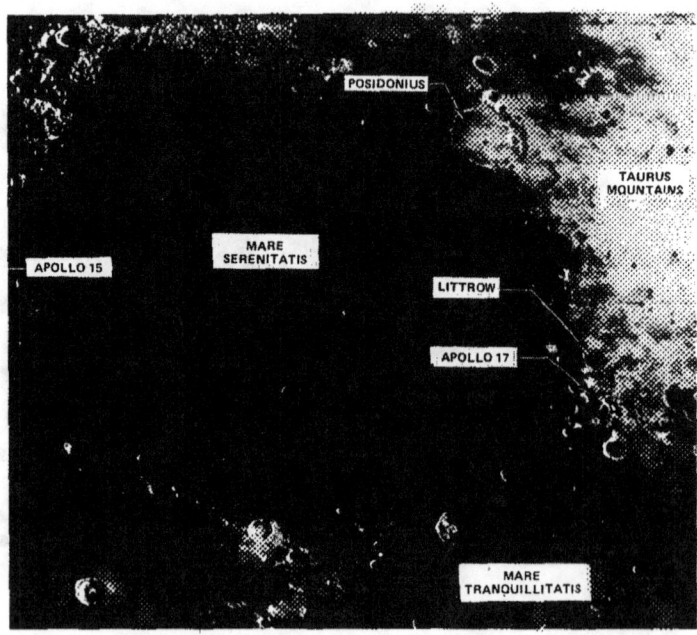

The top photograph is a <u>Lunar Orbiter IV</u> view of <u>Apollo 17</u> landing region. Below, sites that <u>Lunar Orbiter V</u> photographed in August 1967 are plotted on a chart of the Moon's near side. Sites marked S were science and Surveyor sites. Sites marked A were for Apollo. Sites marked X were designated as being of interest for the Apollo Applications Program (the lunar exploration part of Apollo Applications was later cancelled).

D. LUNAR ORBITER PHOTOGRAPHS THE MOON

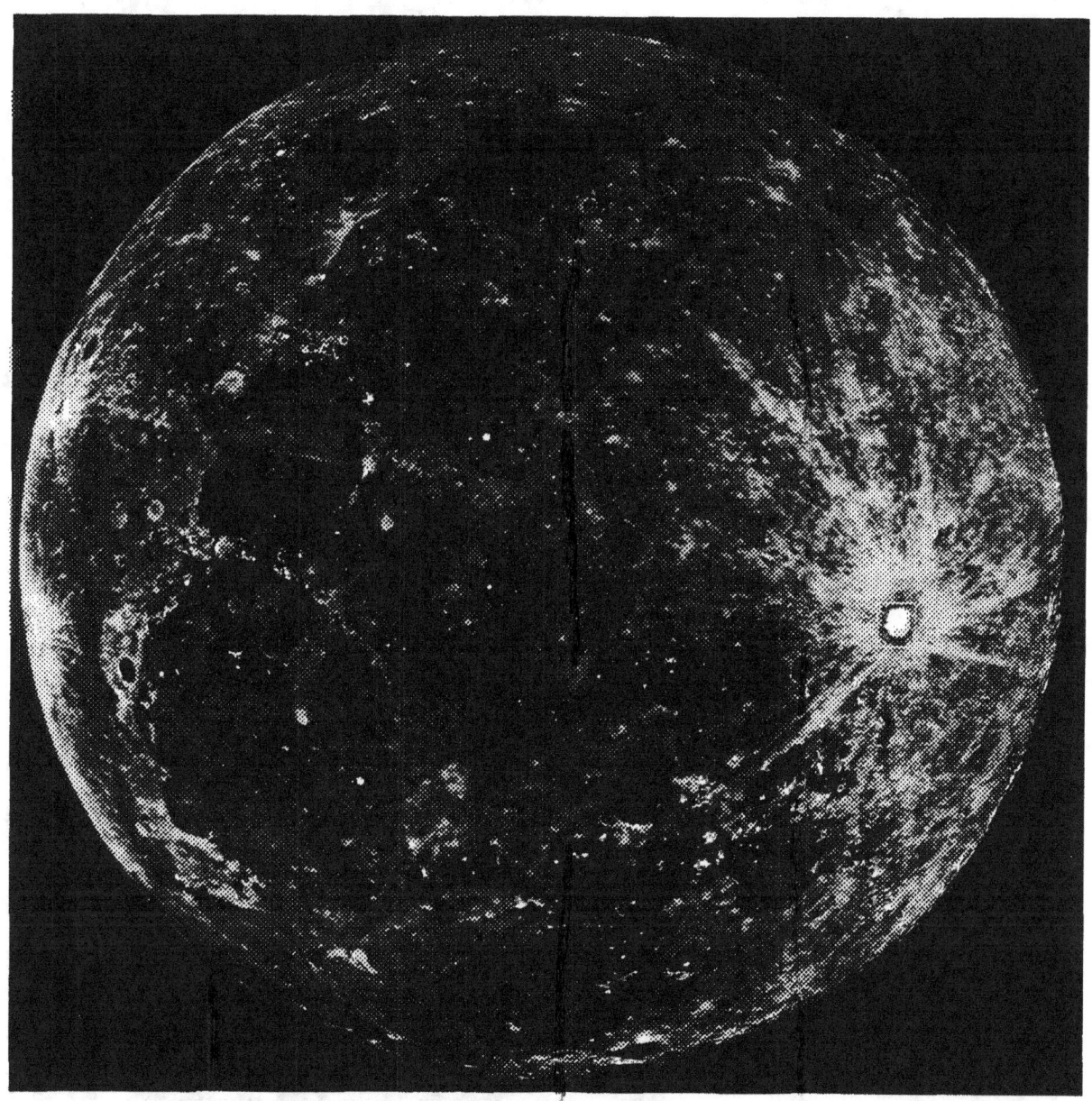

A full view of the Moon photographed from the Lick Observatory, Mount Hamilton, California. The area outlined by the white square is the bright crater Tycho. Two <u>Lunar Orbiter V</u> photos of Tycho follow.

Lunar Orbiter V photographed the 90-kilometer-wide crater Tycho with the wide-angle medium-resolution lens (frame M-123) on August 15, 1967. The view looks almost vertically down onto the crater floor and reveals the central peak, a rough floor, and precipitous walls. The spacecraft was 206 kilometers above the surface of the Moon when this and the following photo were taken.

A high-resolution telephoto picture of part of the floor of the crater Tycho. The area shown is 11.2 by 12.8 kilometers. Fractures, flow markings, and protruding domelike hills with exposed layers suggest a very young floor. The scarcity of smaller impact craters and absence of signs of erosion support the theory that Tycho is a young impact crater.

Copernicus viewed by the 3-meter reflector telescope at Lick Observatory, Mount Hamilton, California, appears as a bull's-eye in this picture. Lunar Orbiter views of this major landmark on the Moon's near side follow.

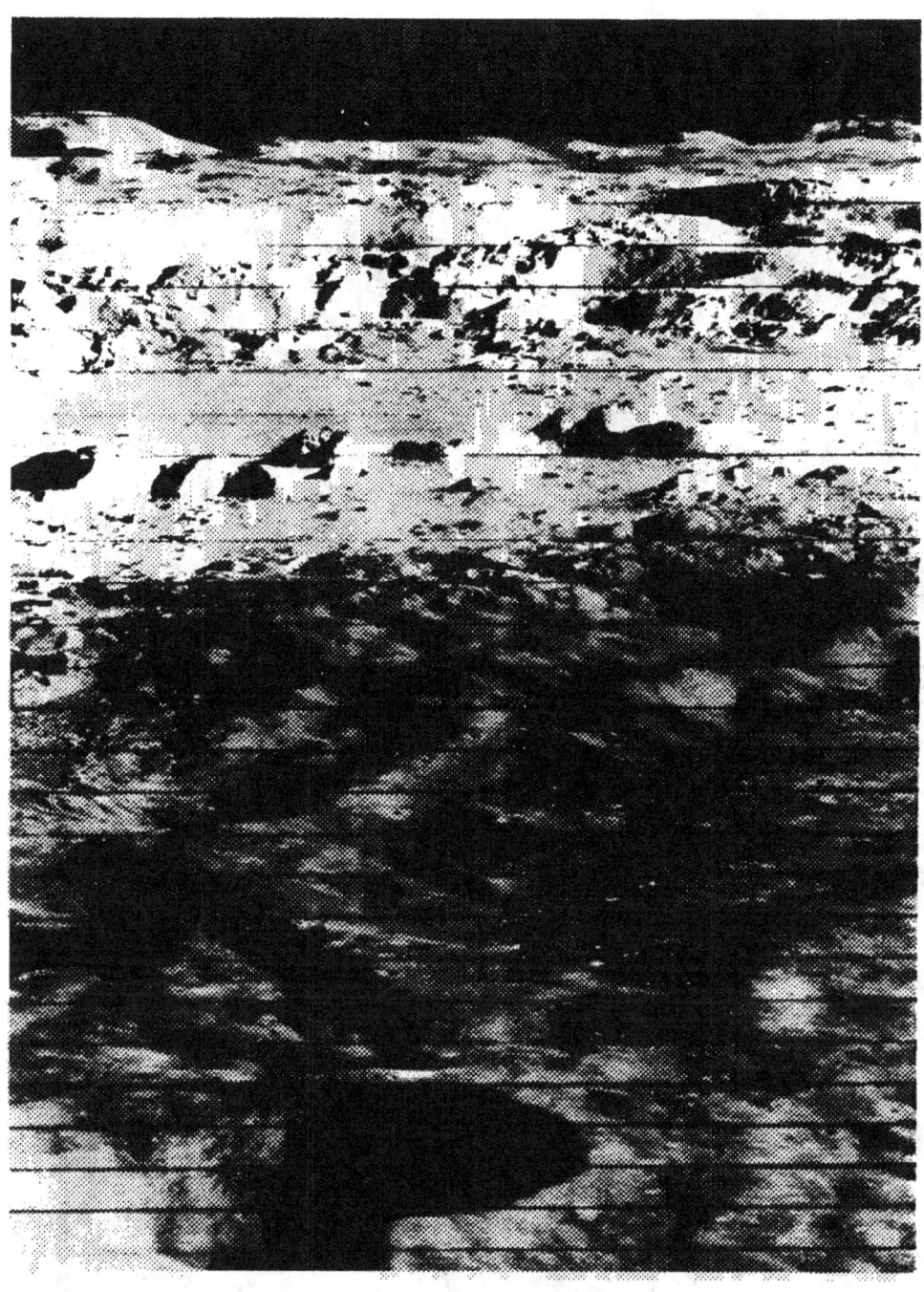

Lunar Orbiter II recorded this oblique view of the crater Copernicus while flying at 43.8 kilometers altitude, 240 kilometers due south of the crater. In the foreground is the "keyhole" crater Fauth, 20.8 kilometers across and 1,372 meters deep. The southern rim of Copernicus is 42.8 kilometers north of Fauth. Copernicus is 96 kilometers in diameter and reaches a depth of 3,200 meters. The Deep Space Network at Goldstone, California, received this picture on November 28, 1966.

An enlargement of the preceding Copernicus photo shows mountains rising 300 meters from the crater floor. Cliffs 300 meters high on the crater rim reveal some downslope movement of material. The horizontal distance across the photograph is about 27.4 kilometers; distance from horizon to the base of the photograph is about 240 kilometers. On the horizon are the Carpathian mountains with the 920-meter-high Gay-Lussac Promontory.

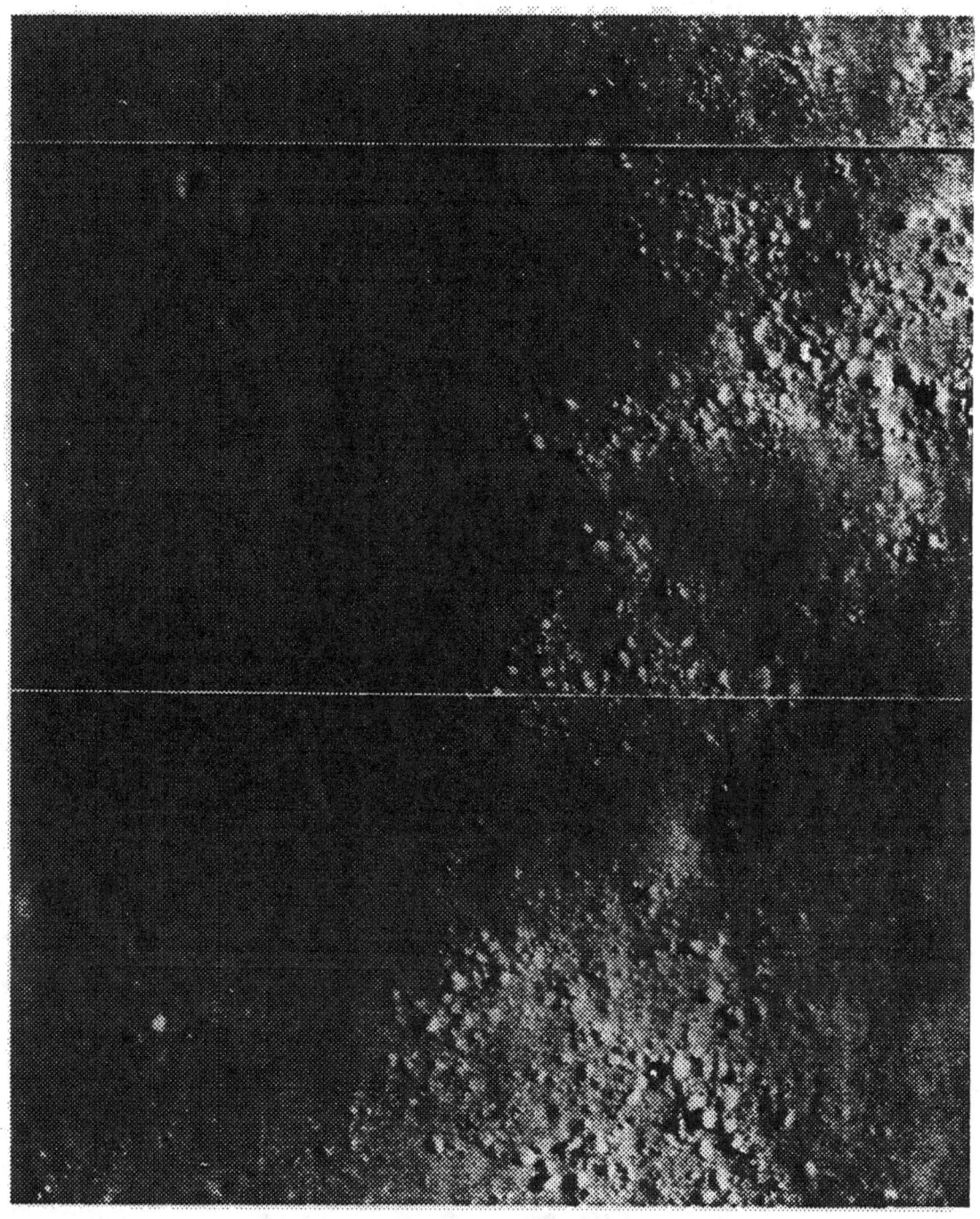

Lunar Orbiter II photographed a rock field in the southeastern part of Mare Tranquillitatis with the 610 mm high-resolution telephoto lens. This picture was enlarged five times from the original film on which the Orbiter photographic data was recorded on Earth. The 365-by 460-meter area is a portion of Site II P-2. Some of the larger rocks in the lower right-hand corner are 10 meters across.

The picture at the left shows the location of the Surveyor I landing site as deduced from horizon features photographed by the Surveyor. Sites I and II seemed compatible with these features. The base map was USAF Aeronautical Charting and Information Center's Lunar Chart LAC 75.

The three photos opposite, taken by Lunar Orbiter III February 22, 1967, enabled NASA to pinpoint the location of Surveyor I. The left photo is of the area north of the crater Flamsteed, where the Surveyor landed June 2, 1966. The black lines point to low mountains photographed by the Surveyor. The center photo is a vertical view of the area outlined in the black rectangle in the oblique picture to the left. The square in the center photo encloses the area of the Surveyor landing site that is pictured greatly enlarged at the right. The magnitude of the light reflected from Surveyor I, the long pointed shadow, and the triangulation of Orbiter and Surveyor photos confirmed this as the landed spacecraft.

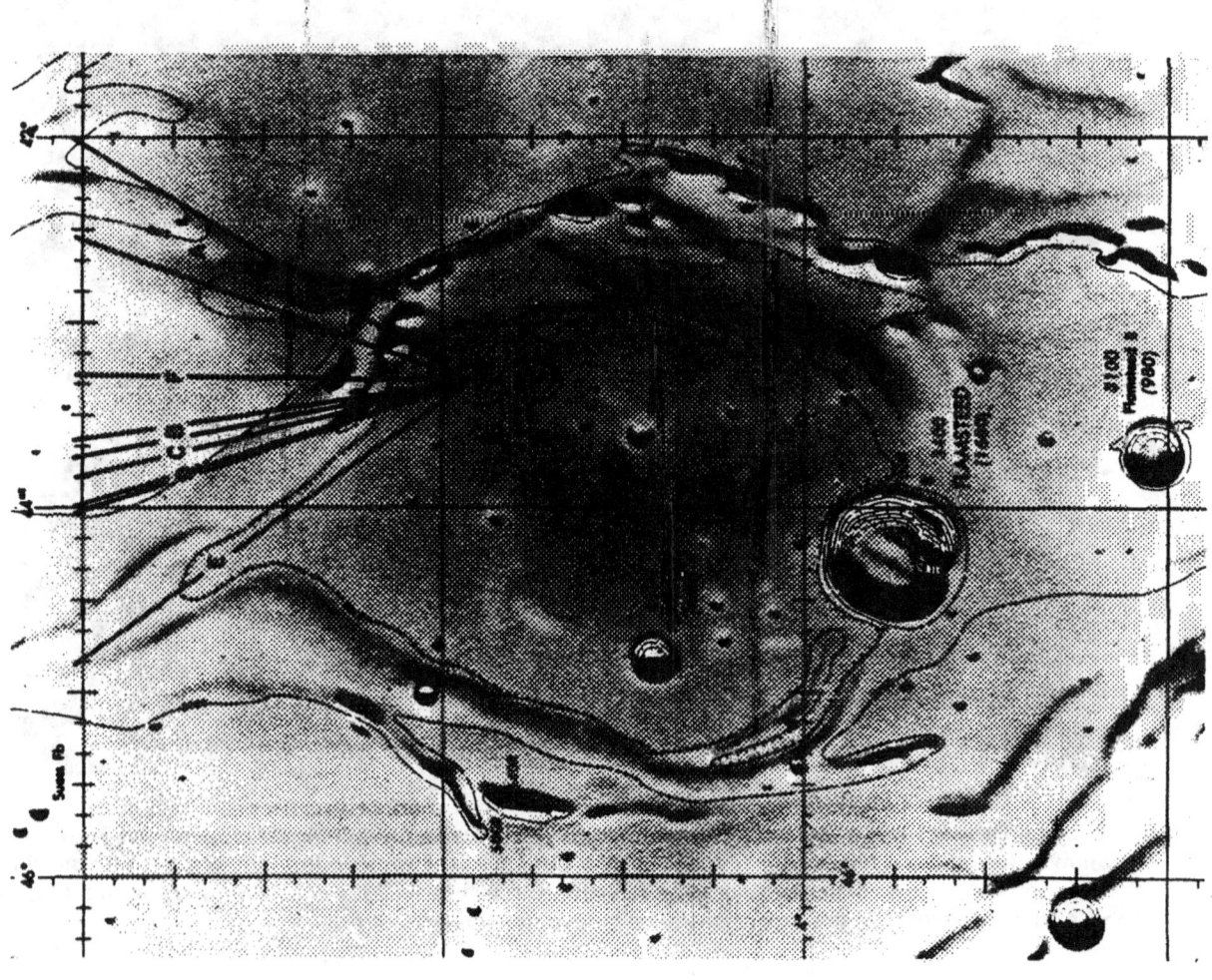

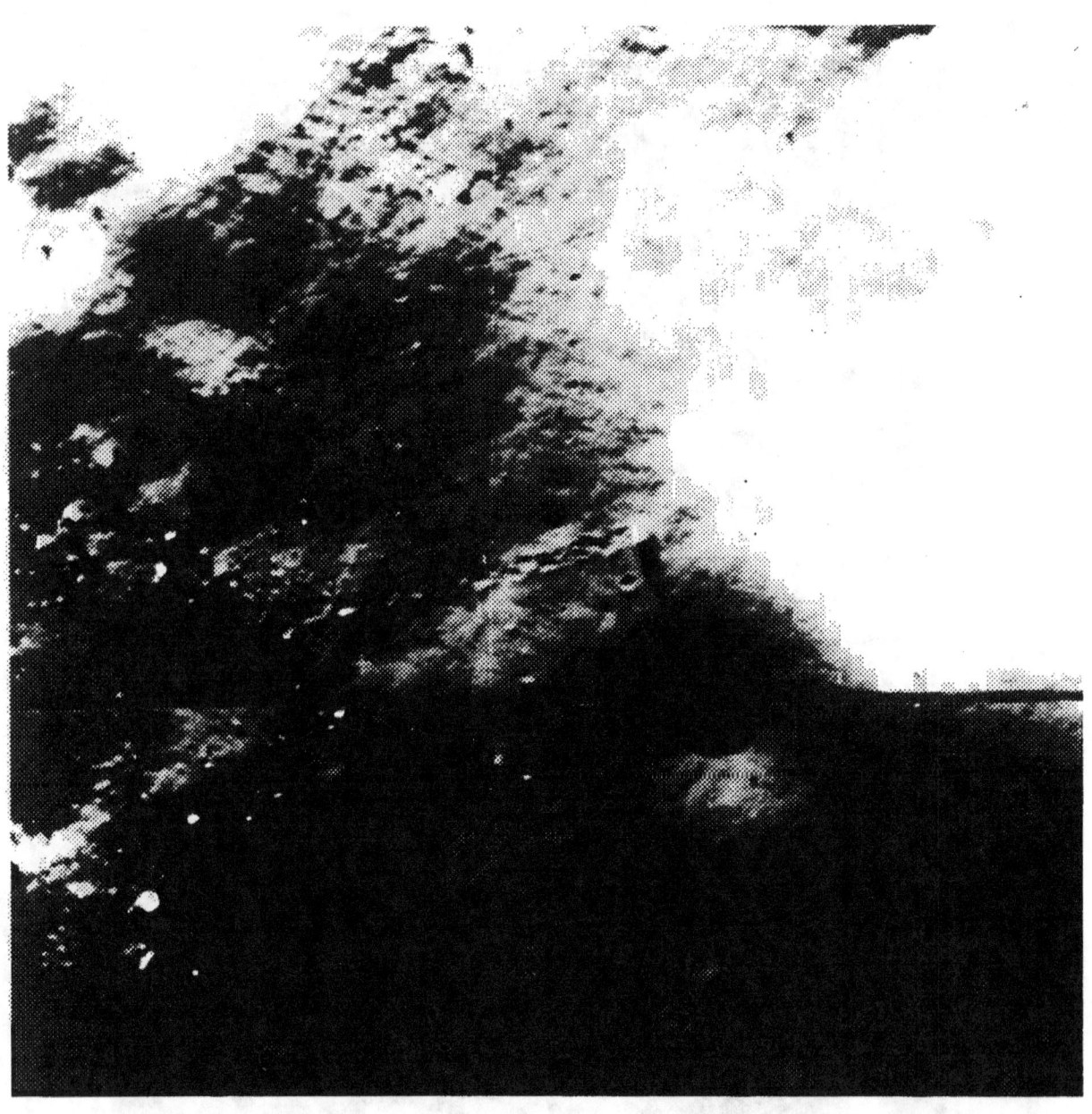

Lunar Orbiter V photographed an area in the Vitello crater (south of Mare Humorum at 30.61° S latitude, 37.57° W longitude) on August 17, 1967. The enlarged portion of that high-resolution telephoto picture reveals two large "rolling stones," whose paths are clearly visible. The larger one near the center of the picture is about 23 meters across and has rolled or bounced some 274 meters. The smaller rock is 4.6 meters across and has traveled 365 meters. Numerous boulder tracks in Orbiter pictures have told scientists much about the soil mechanics of the lunar surface, its cohesiveness and bearing strength, and the possibility of quakes as one cause of rock movement on the Moon.

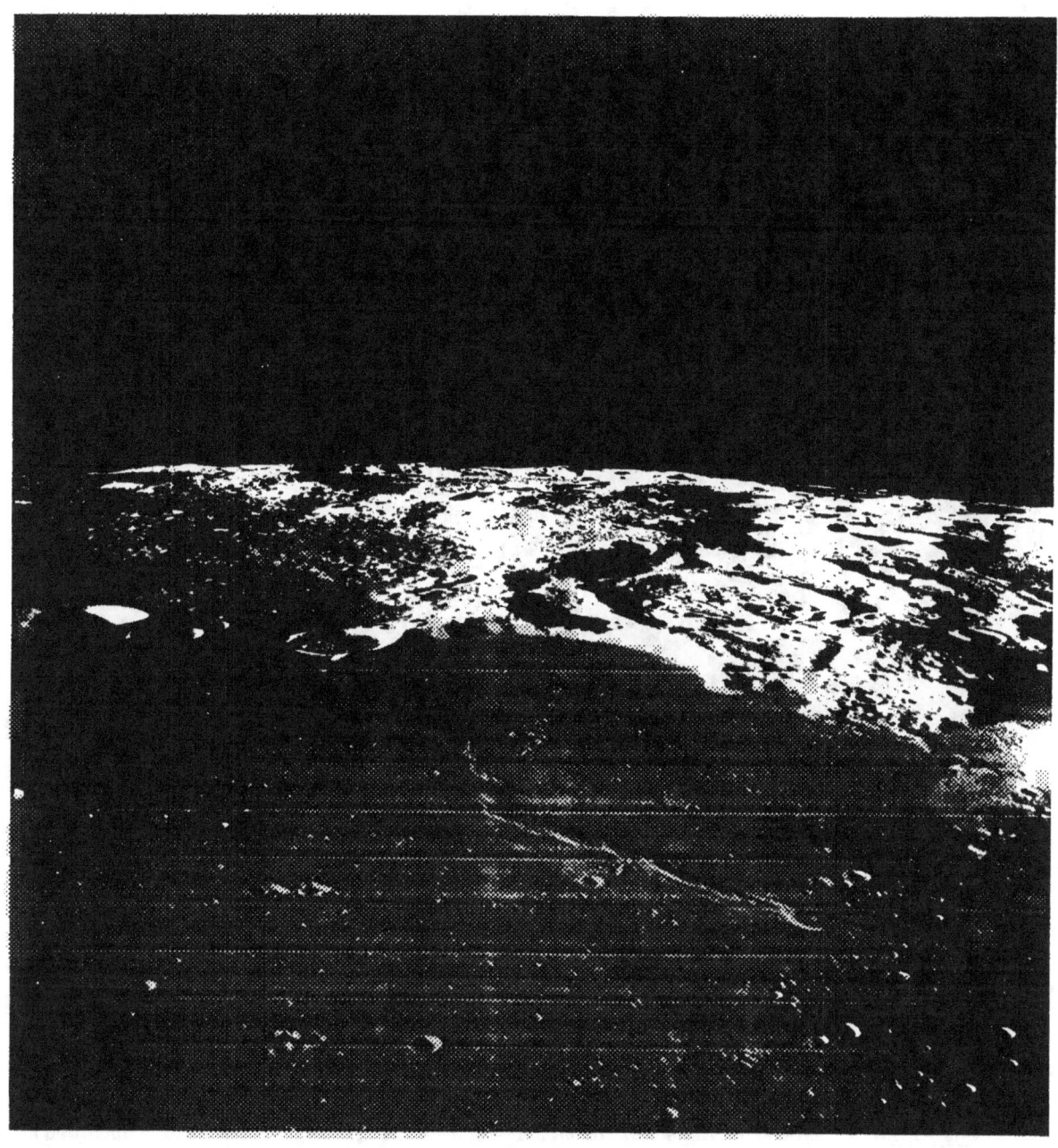

From an altitude of only 56 kilometers <u>Lunar Orbiter III</u> photographed the crater Damoiseau and surrounding area in the southwestern part of Oceanus Procellarum on February 22, 1967. The inner crater is 40 kilometers in diameter and the outer crater 56 kilometers. The crater resembles a geological phenomenon known on Earth as a caldera, a volcanic structure including an area of collapsed material. The contact between mare floor and upland areas is sharply defined here. Damoiseau was scheduled as Science site S-29 on the third Orbiter mission. The picture is from frame M-213.

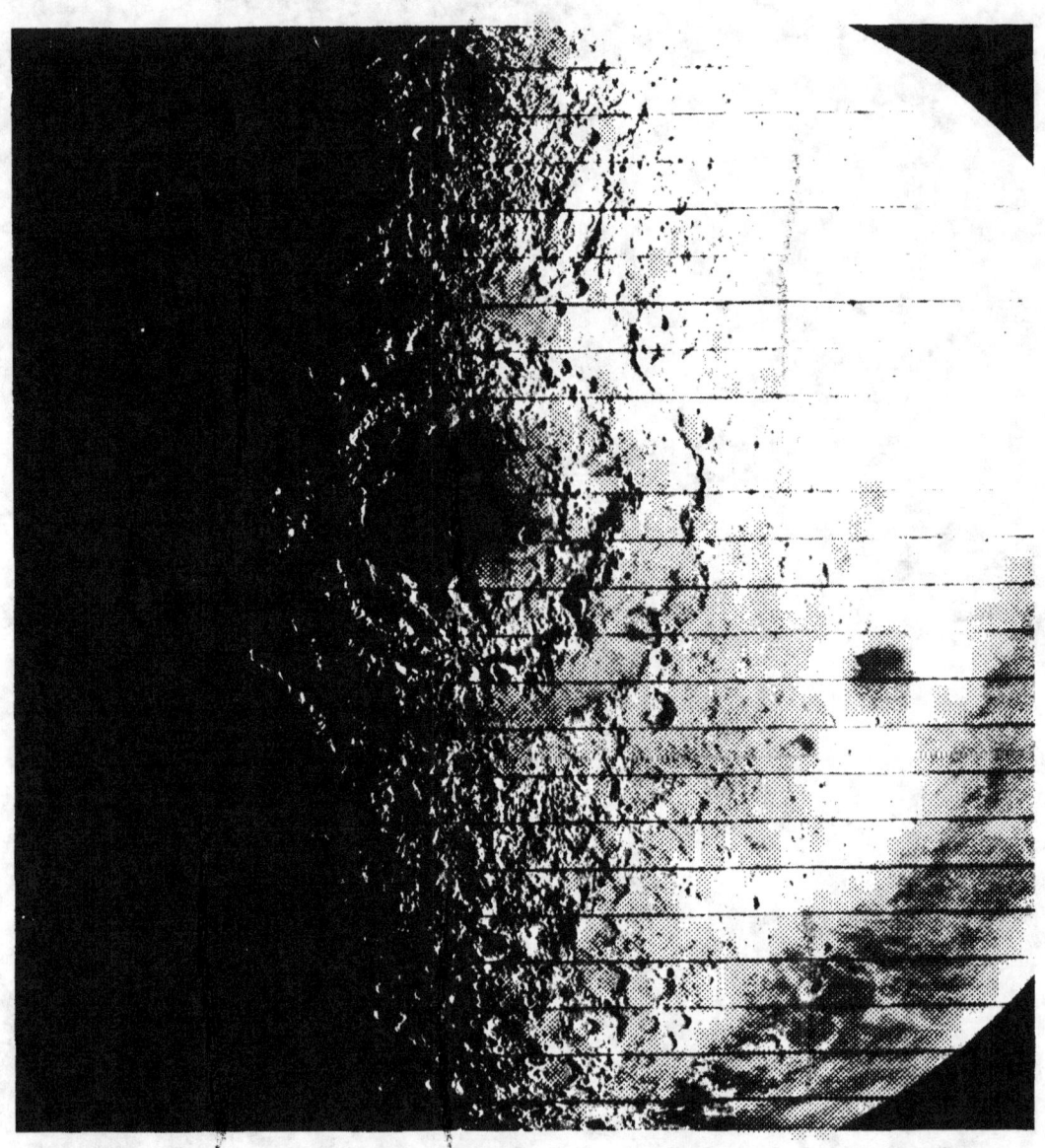

Lunar Orbiter IV wide-angle frame 187, taken May 25, 1967, at 2,720 kilometer altitude, recorded an enormous, complex feature on the lunar surface, the Orientale Basin. Centered at 89° W longitude, 15° S latitude, the gigantic circular basin measures 965 kilometers in diameter at its outer scarp. At this perimeter the Cordillera Mountains, ringing the basin, rise 6,100 meters and are the most massive on the Moon. Within the outer ring the Rook Mountains form another circular scarp about 640 kilometers in diameter. Surrounding this complex basin, a coarsely graded blanket extends another 965 kilometers over the older cratered surface. The freshness of the surface texture and sharpness of the mountain areas suggest that Orientale is among the youngest large circular basins on the Moon. If it and the surrounding scarps and blanket were formed by a meteorite impact, as seems possible, then a high-velocity body of asteroidal proportions struck the Moon with monumental force.

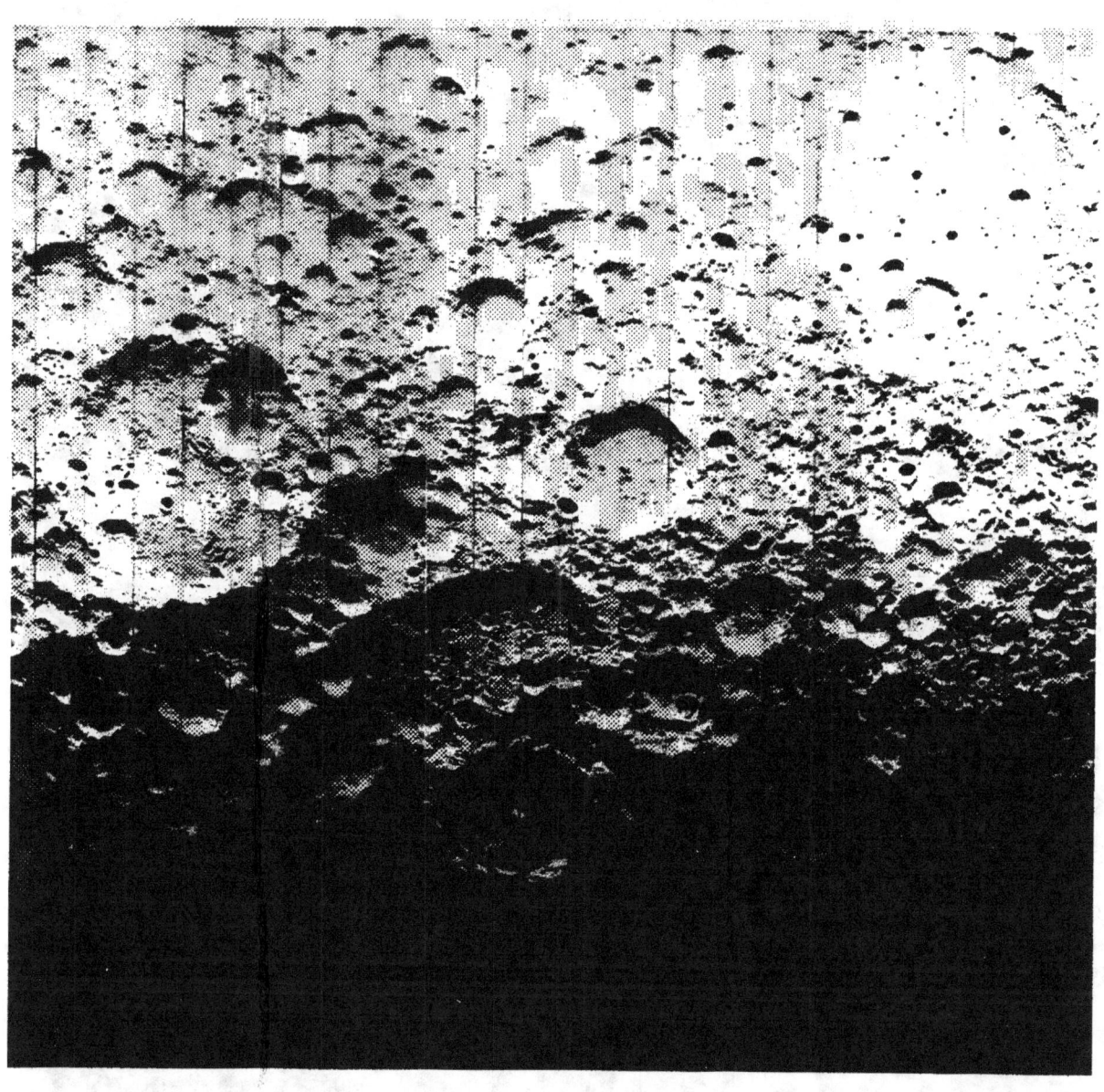

Lunar Orbiter I recorded this view of the Moon's heavily cratered far side in frame 116 on August 24, 1966. The area shown here covers a rectangle 1,300 by 1,450 kilometers and is in the eastern portion of the far side just at the terminator as viewed from Earth.

On August 10, 1967, <u>Lunar Orbiter V</u> made this wide-angle westward-looking oblique photo of the elongated crater Messier and Messier A. These craters are at 47° E longitude, 2° S latitude on the floor of Mare Fecunditatis. A double ray from the pair of craters extends westward for about 160 kilometers. One interpretation for the peculiar shape of Messier and the rays is that they may have been produced by a low-angle impact of a meteorite on the Moon.

Lunar Orbiter V telephoto frame 41 shows details of craters Messier and Messier A. Taken at an altitude of 97 kilometers, the picture reveals features on the floor of Messier, which is about 13 kilometers long, 10 kilometers wide, and 1,220 meters deep. Material ejected from the craters can be seen on the mare floor. Downslope movement of material in Messier shows some accumulation on the crater floor. Messier A is 13 kilometers across.

E. LUNAR ORBITER PHOTOGRAPHS THE EARTH

The first Earth photograph by <u>Lunar Orbiter I</u> shows the cratered lunar horizon and the swirling cloud masses on Earth some 345,700 kilometers away. Taken August 23, 1966, as the spacecraft was about to pass behind the Moon on its 16th orbit, the picture proved valuable to program scientists for what it showed of the lunar surface at an oblique rather than a vertical angle. The illuminated crescent of the Earth shows the U. S. East Coast in the upper left, southern Europe toward the night side of Earth and Antarctica at the bottom of the crescent.

On August 8, 1967, <u>Lunar Orbiter V</u> took this photo of the nearly full Earth with the 610 mm lens. The exposure time was 1/100 second, which was insufficient to compensate for the Earth's high albedo (about 0.36 of 1.0). However, ground processing successfully compensated for overexposure. The subsolar point was just above and left of the Aral Sea, and the spacecraft's camera line of sight with Earth focused on a point slightly above and right of the Aral Sea. The angle between the subsolar point and the camera's line-of-sight axis intercept was 31.5°. The spacecraft was about 5,860 kilometers above the Moon in near polar orbit, so that the surface is not seen. The picture shows Italy, Greece, Turkey, the Mediterranean, the Red Sea, most of the African continent, Madagascar, India, and Central Asia.

F. PROGRAM MANAGEMENT

Dr. Floyd L. Thompson (above, left), Director of Langley Research Center at the time of the Lunar Orbiter Program, and Capt. Lee R. Scherer, Lunar Orbiter Program Manager at NASA Headquarters, discussed final mission results October 17, 1967. Behind them is a partial mosaic of the Moon's surface made from Lunar Orbiter photos. Kneeling on the "Moon" (left) Langley Lunar Orbiter Project Manager Clifford H. Nelson examined a section of the 1-meter-square mosaic of 127 <u>Lunar Orbiter IV</u> photos. The U. S. Army Map Service assembled the mosaic for Langley Research Center.

Oran W. Nicks (above, left), NASA Director of Lunar and Planetary Programs, and Robert J. Helberg, Boeing's Orbiter Program Manager, watched thermal shroud fittings in 1965. Below, the mission monitoring group during Lunar Orbiter II's November 1966 mission included NASA Program Director Scherer (standing at left) and (left to right) Neil A. Holmberg, A. Thomas Young, Uriel M. Lovelace, Leon J. Kosofsky, Joseph Brenkle (standing), Dr. Thor Karlstrom, and Gerald W. Brewer.

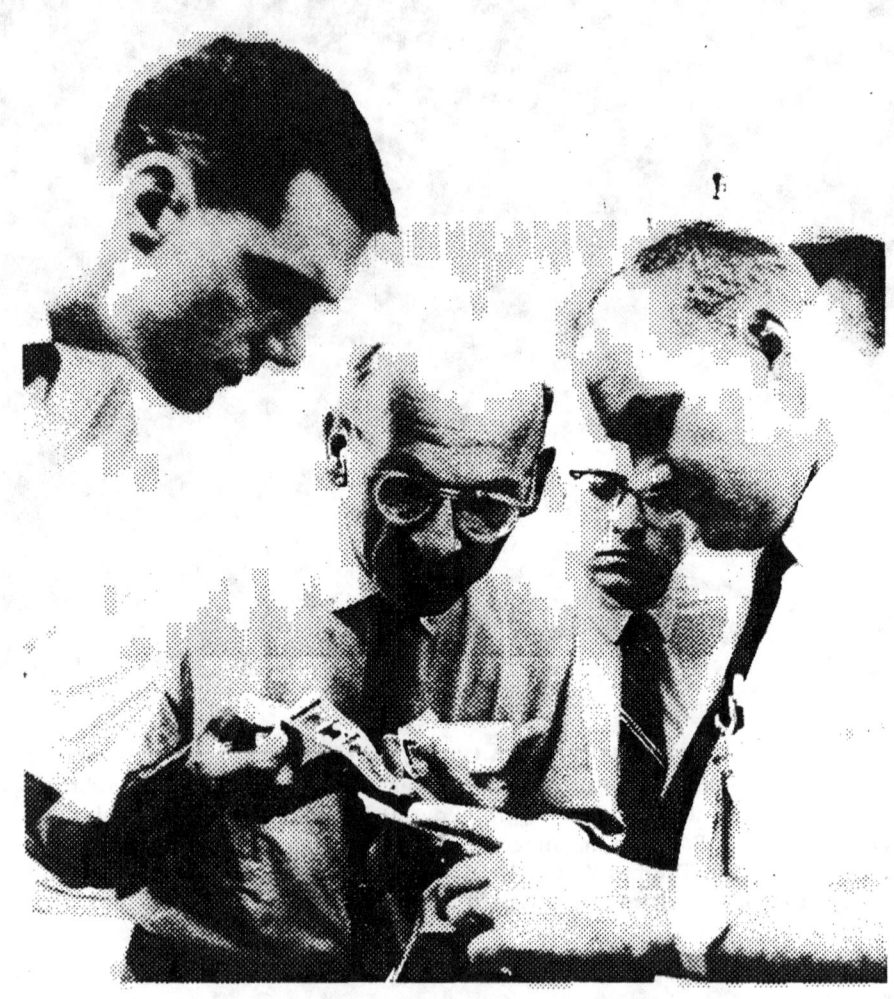

Israel Taback (center), Lunar Orbiter Spacecraft Manager from Langley Research Center, examined a reconstructed photograph from <u>Lunar Orbiter II</u> with John B. Graham of Operations Integration (right). Picture data from the spacecraft were received at the Deep Space Network Tracking Station at Goldstone, California, and routed to photographic ground reconstruction equipment at the Jet Propulsion Laboratory's Space Flight Operations Center in Pasadena.

Appendix A

GLOSSARY

albedo -- The ratio of the amount of electromagnetic radiation reflected by a body to the amount incident upon it. This concept is identical with reflectance but should be distinguished from spectral reflectance.

anomaly -- In general, a deviation from the norm, an irregularity, a malfunction.

apolune -- That point in a lunar-centric orbit which is most distant from the Moon.

Bimat web -- The continuous processing film used in the Lunar Orbiter photographic subsystem to process the camera film.

deboost -- A velocity control engine burn to allow a spacecraft to decelerate and go into orbit around a planetary body, or to leave an orbit and descend to a landing on that body.

delta V -- A change in velocity.

flux -- The rate of flow of some quantity, often used in reference to the flow of some form of energy.

gimbal -- A device with two mutually perpendicular and intersecting axes of rotation. It provides free angular movement in two directions and serves as an engine mount.

Greenwich Mean Time (GMT) -- The local mean time at the Greenwich, England, meridian. Some of the Lunar Orbiter post-launch operations reports used the local time at the Kennedy Space Center, expressed either in Eastern Standard Time (EST, 5 hours behind GMT) or Eastern Daylight Time (EDT, 4 hours behind GMT), depending on the time of year when a launch took place.

ground resolution -- The degree to which an optical or photographic system can reproduce fine detail of the surface being imaged, as measured against a photographic scale. It is the product of a combination of capabilities of the film (graininess, sensitivity, etc.) and the lens (type, resolving power, etc.) and is usually expressed in line pairs per millimeter. Photographic scale is found by dividing the altitude at which the picture is taken by the focal length of the camera. For Lunar Orbiter, the effective film resolution was 76 line pairs per millimeter, which gave 1-meter resolution through the 610 mm lens and 8-meter resolution through the 80 mm lens under predetermined contrast conditions on the lunar surface.

hypergolic -- A term used to describe propellants that ignite spontaneously on contact with an oxidizer; a self-igniting fuel, propellant, or propulsion system.

launch window -- The postulated opening in a continuum of time or space through which a spacecraft must be launched to achieve a desired encounter, rendezvous, or impact.

noise level -- The level of any undesired disturbance within a useful frequency band.

nominal -- Occurring or performing as intended in pre-mission planning.

oxidizer -- A substance that combines with another to produce heat and, in a rocket, hot gases of combustion thrust.

parking orbit -- A temporary orbit in which a vehicle coasts before transfer into final orbit or trajectory.

perilune -- The point in a lunar-centric orbit which is closest to the Moon.

pitch -- An angular movement (of a spacecraft) about an axis parallel to the lateral axis of the vehicle.

roll -- The rotational or oscillatory movement of a spacecraft or similar body about a longitudinal axis through the spacecraft.

software -- (Computer) programs and formulation of programs.

yaw -- The rotational or oscillatory movement of a spacecraft or the like about a vertical axis.

Sources: Charles McLaughlin, *Space Age Dictionary* (Princeton: D. Van Nostrand Co., 1959). William H. Allen, ed., *Dictionary of Technical Terms for Aerospace Use*, NASA SP-7 (Washington, D.C.: GPO, 1965). J. L. Nayler, *A Dictionary of Astronautics* (New York: Hart Publishing Co., 1964). Woodford A. Heflin, ed., *The United States Air Force Dictionary* (Air University Press, 1956).

APPENDIX B
ORGANIZATION CHARTS

NASA HEADQUARTERS LUNAR ORBITER PROGRAM ORGANIZATION

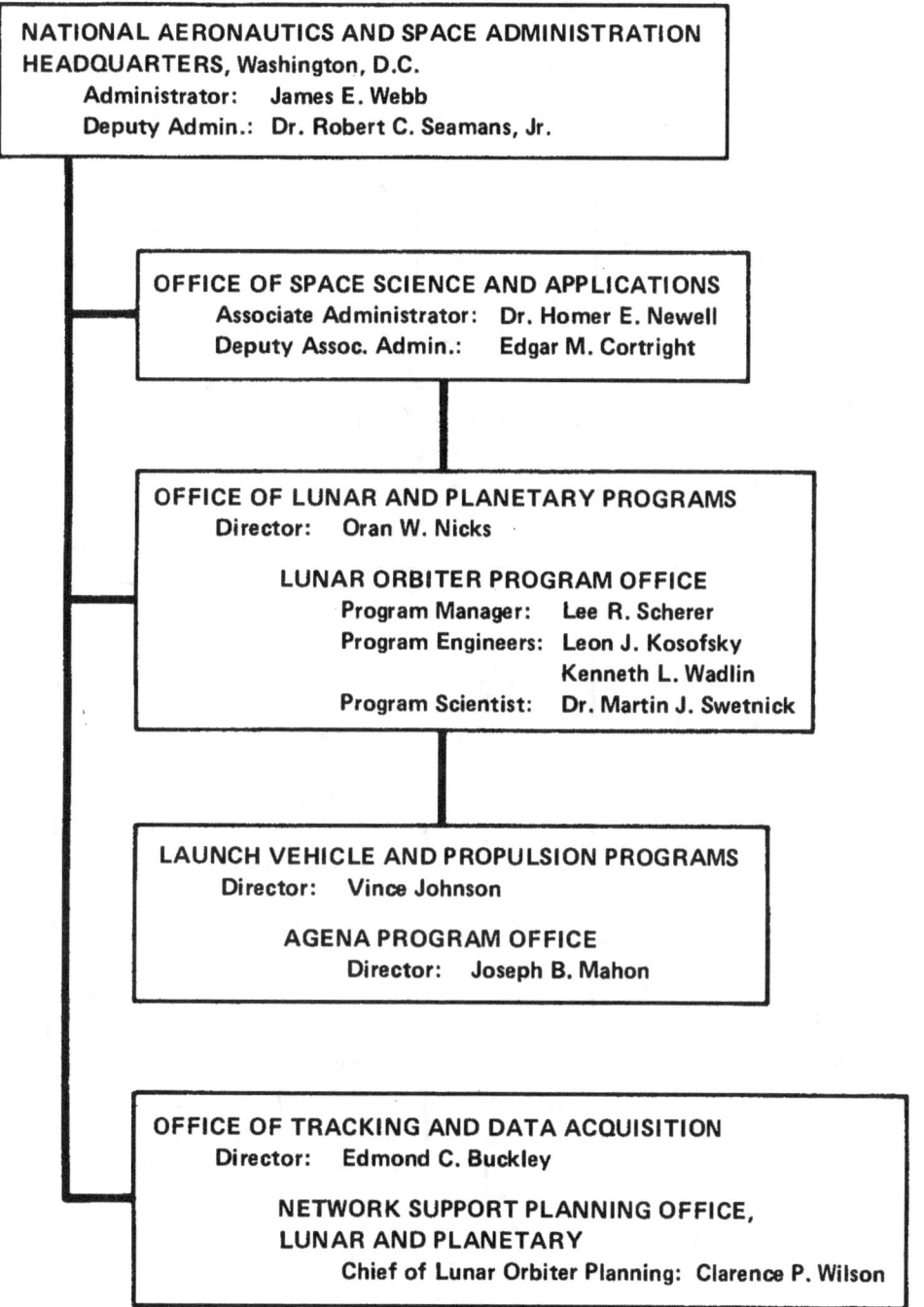

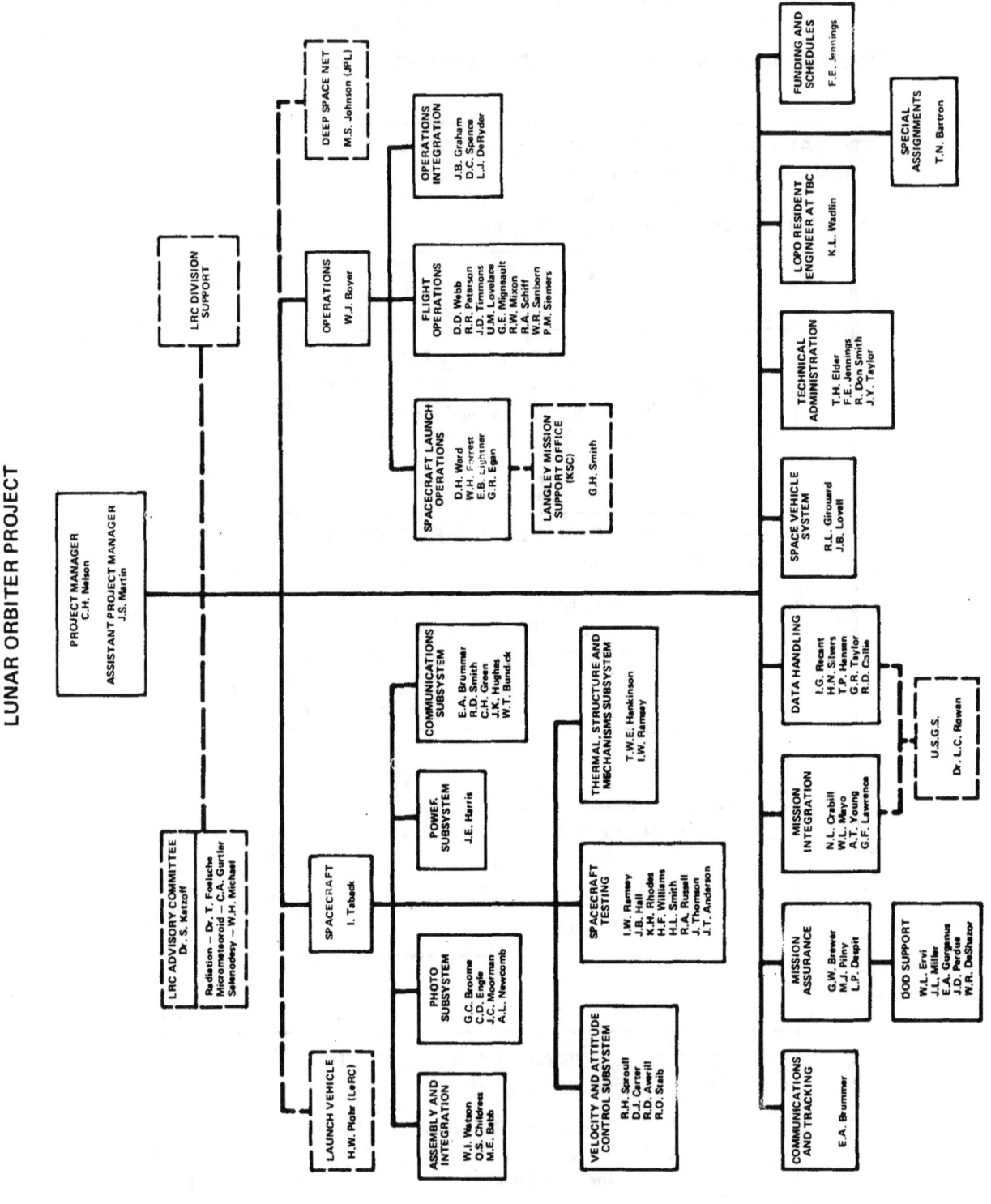

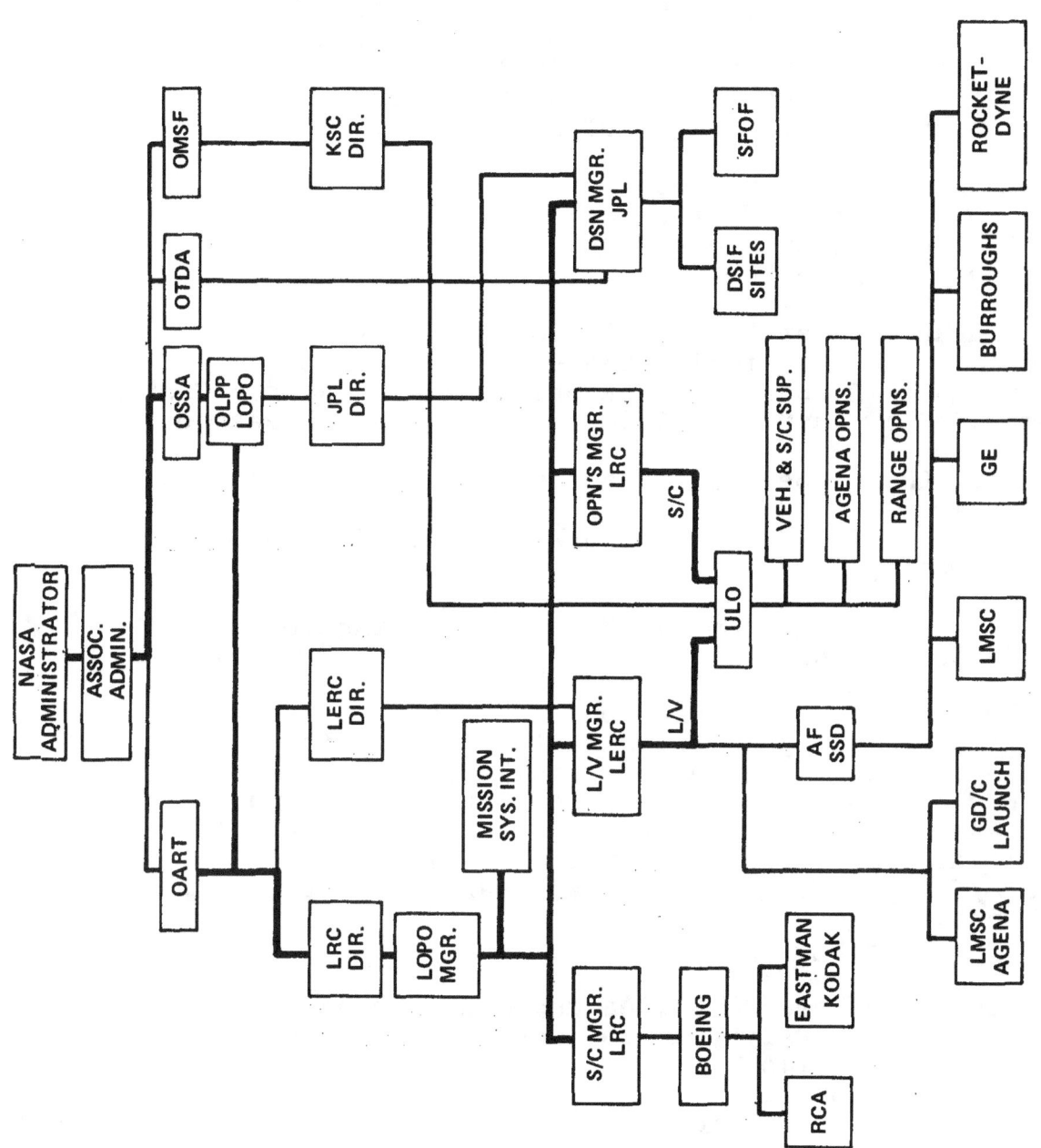

OVERALL PROJECT MANAGEMENT ORGANIZATION FOR LUNAR ORBITER*

*The key to this chart is on the next page.
Source: Project Development Plan, Lunar Orbiter Project. Project No. 814-00-00, December 10, 1964, Revised June 10, 1966, prepared by the Langley Research Center, Langley Station, Hampton, Virginia, p. V-3.

Key to Overall Project Management Chart

NASA Administrator: James E. Webb
Assoc. Admin.: Dr. Robert C. Seamans, Jr.
OART: Office of Advanced Research and Technology
OSSA: Office of Space Science and Applications
OTDA: Office of Tracking and Data Acquisition
OMSF: Office of Manned Space Flight
OLPP: Office of Lunar and Planetary Programs
LOPO: Lunar Orbiter Program Office

LRC: Langley Research Center
LeRC: Lewis Research Center
JPL: Jet Propulsion Laboratory
KSC: Kennedy Space Center

LOPO: Lunar Orbiter Project Office, Manager
Mission Sys. Int.: Mission Systems Integration
S/C Mgr. LRC: Spacecraft Manager, Langley Research Center
L/V Mgr. LeRC: Launch Vehicle Manager, Lewis Research Center
Opn's Mgr. LRC: Operations Manager, Langley Research Center
DSN Mgr. JPL: Deep Space Network Manager, Jet Propulsion Laboratory

Boeing: The Boeing Company, Seattle, Washington
L/V: Launch Vehicle
S/C: Spacecraft
ULO: Unmanned Launch Operations
DSIF Sites: Deep Space Instrumentation Facility Sites
SFOF: Space Flight Operations Facility
RCA: Radio Corporation of America, Princeton, N.J.
Eastman Kodak: Rochester, N.Y.
AF SSD: Air Force Support Services Division
Veh. & S/C Sup. Vehicle and Spacecraft Support
Agena Opns.: Agena Operations
Range Opns.: Range Operations

LMSC Agena: Lockheed Missiles and Space Company
GD/C Launch: General Dynamics, Convair Division
GE: General Electric
Burroughs: subcontractor
Rocketdyne: subcontractor

Appendix C

RECORD OF UNMANNED LUNAR PROBES, 1958-1968

I. United States

Probe	Launch Date (local time)	Launch Site	Launch Vehicle	Spacecraft Weight (kg)	Mission Results
Pioneer I	Oct. 11, 1958	ETR	Thor-Able	38	Reached 113,783-km altitude before disintegrating in Earth's atmosphere Oct. 12; insufficient final velocity.
Pioneer III	Dec. 6, 1958	ETR	Juno II	6	Reached 102,322-km altitude before disintegrating in Earth's atmosphere Dec. 7; insufficient final velocity.
Pioneer IV	Mar. 3, 1959	ETR	Juno II	6	Passed within 60,000 km of Moon and went into solar orbit.
Ranger I	Aug. 23, 1961	ETR	Atlas-Agena B	306	Disintegrated Aug. 30 on failure to achieve intended Earth orbit. Orbit too low.
Ranger II	Nov. 18, 1961	ETR	Atlas-Agena B	306	Disintegrated Nov. 18 after failing to achieve Earth orbit.
Ranger III	Jan. 26, 1962	ETR	Atlas-Agena B	330	Missed Moon by 36,790 km; went into solar orbit.

Probe	Launch Date (local time)	Launch Site	Launch Vehicle	Spacecraft Weight (kg)	Mission Results
Ranger IV	Apr. 23, 1962	ETR	Atlas-Agena B	331	Mission unsuccessful because camera equipment failed to function; faulty programmer. First U.S. spacecraft to hit Moon.
Ranger V	Oct. 18, 1962	ETR	Atlas-Agena B	342	Mid-course correction failed; spacecraft missed Moon by 720 km, went into solar orbit.
Ranger VI	Jan. 30, 1964	ETR	Atlas-Agena B	365	Precise lunar impact. Photographic mission unsuccessful because premature turn-on caused camera failure.
Ranger VII	Jul. 28, 1964	ETR	Atlas-Agena B	366	Successfully sent back 4,316 high-resolution TV photos during last 13 min of flight before precise impact on Moon.
Ranger VIII	Feb. 17, 1965	ETR	Atlas-Agena B	367	Transmitted 7,137 close-up TV photos of Moon before precise impact in Sea of Tranquility.
Ranger IX	Mar. 21, 1965	ETR	Atlas-Agena B	367	Transmitted 5,814 TV photos before precise impact in crater Alphonsus. First high-resolution photos of lunar crater.
Surveyor I	May 30, 1966	ETR	Atlas-Centaur	270	Softlanded on Moon; transmitted 11,237 TV photos in 13 days; survived one lunar night.

Probe	Launch Date (local time)	Launch Site	Launch Vehicle	Spacecraft Weight (kg)	Mission Results
Explorer XXXIII (IMP-D)	Jul. 1, 1966	ETR	Thrust-augmented Thor-Delta	93	Failed to achieve lunar orbit because of launch errors; remained in Earth orbit.
Lunar Orbiter I	Aug. 10, 1966	ETR	Atlas-Agena D	386	Entered lunar orbit Aug. 14. Photographed Moon until Aug. 29, photographing all 9 primary and 7 potential Apollo sites, 11 areas on far side. Impacted Moon Oct. 29, 1966.
Surveyor II	Sep. 20, 1966	ETR	Atlas-Centaur	270	Vernier failed. Spacecraft crashed on Moon SE of crater Copernicus.
Lunar Orbiter II	Nov. 6, 1966	ETR	Atlas-Agena D	385	Returned 205 lunar frames, including 13 primary and 17 secondary Apollo sites. Impacted surface of Moon Oct. 11, 1967.
Lunar Orbiter III	Feb. 4, 1967	ETR	Atlas-Agena D	386	Photographed Surveyor I on Moon. Returned 182 lunar frames. Impacted Moon Oct. 9, 1967.
Surveyor III	Apr. 17, 1967	ETR	Atlas-Centaur	281	Softlanded on Moon Apr. 19; soil sampler, photo experiments until May 3, 1967. Took 6,315 photos.
Lunar Orbiter IV	May 4, 1967	ETR	Atlas-Agena D	386	Returned 163 frames; impacted Moon Oct. 6, 1967.

Probe	Launch Date (local time)	Launch Site	Launch Vehicle	Spacecraft Weight (kg)	Mission Results
Surveyor IV	Jul. 14, 1967	ETR	Atlas-Centaur	281	Signal lost 2.5 min before landing on Moon July 17.
Lunar Orbiter V	Aug. 1, 1967	ETR	Atlas-Agena D	386	Last mission of Lunar Orbiter photo-mapping program. Returned 212 frames; photographed 23 new sites on far side, 1st "full earth" view, 36 scientific sites, 5 Apollo sites. Impacted Moon Jan. 31, 1968.
Surveyor V	Sep. 8, 1967	ETR	Atlas-Centaur	281	Softlanded on Moon Sep. 10. Returned over 19,000 photos, soil analysis data.
Surveyor VI	Nov. 7, 1967	ETR	Atlas-Centaur	282	Softlanded on Moon Nov. 9. Performed 1st rocket liftoff from lunar surface, moving 2.5 m. Analyzed soil 27 hrs. Transmitted 30,065 TV photos.
Surveyor VII	Jan. 7, 1968	ETR	Atlas-Centaur	290	Softlanded on Moon Jan. 10. Transmitted over 21,000 TV photos. Analyzed soil and dug trench. Photographed Earth and Jupiter. Surveyor program ended with 5th success in 7 tries.

II. Soviet Union

Probe	Launch Date (local time)	Launch Site	Launch Vehicle	Spacecraft Weight (kg)	Mission Results
Luna I	Jan. 2, 1959	Tyuratam	A-1	361	Passed within 6,000 km of Moon; went into solar orbit.
Luna II	Sep. 12, 1959	Tyuratam	A-1	390	Struck Moon Sep. 13; 1st man-made device to reach another celestial body.
Luna III	Oct. 4, 1959	Tyuratam	A-1	435	Recorded 1st photographic data on Moon's far side. Reentered Earth's atmosphere Apr. 20, 1960.
Luna IV	Apr. 2, 1963	Tyuratam	A-2-e	1,422	Passed within 8,500 km of lunar surface.
Luna V	May 9, 1965	Tyuratam	A-2-e	1,476	Struck lunar surface in unsuccessful softlanding attempt May 12.
Luna VI	Jun. 8, 1965	Tyuratam	A-2-e	1,442	Intended for softlanding; missed Moon by 160,000 km, went into solar orbit.
Zond III	Jul. 18, 1965	Tyuratam	A-2-e	960	Flew by Moon and sent back photographic data; went into solar orbit.
Luna VII	Oct. 4, 1965	Tyuratam	A-2-e	1,506	Intended for lunar softlanding. Retrorockets fired prematurely, causing impact on Moon Oct. 7.

Probe	Launch Date (local time)	Launch Site	Launch Vehicle	Spacecraft Weight (kg)	Mission Results
Luna VIII	Dec. 3, 1965	Tyuratam	A-2-e	1,552	Intended for lunar softlanding. Retrorockets fired late, causing spacecraft to impact Moon Dec. 7.
Luna IX	Jan. 31, 1966	Tyuratam	A-2-e	1,583	First spacecraft to softland on Moon (Feb. 3) and transmit TV pictures of landing site to Earth.
Luna X	Mar. 31, 1966	Tyuratam	A-2-e	1,600	First spacecraft to orbit Moon. Studied micrometeoroid flux, lunar environment until May 30, 1966.
Luna XI	Aug. 24, 1966	Tyuratam	A-2-e	1,640	Entered lunar orbit Aug. 27. Sent back data until Oct. 1, 1966.
Luna XII	Oct. 22, 1966	Tyuratam	A-2-e	1,670?	Entered lunar orbit Oct. 25. Studied radiation, transmitted photos of lunar surface.
Luna XIII	Dec. 21, 1966	Tyuratam	A-2-e	1,670?	Softlanded on Moon Dec. 24. Tested hardness of lunar surface, photographed lunar panorama.
Luna XIV	Apr. 7, 1968	Tyuratam	A-2-e	1,670?	Entered lunar orbit Apr. 10. Studied Earth-Moon mass relationships, Moon's gravitational field; no photos returned.

Probe	Launch Date (local time)	Launch Site	Launch Vehicle	Spacecraft Weight (kg)	Mission Results
Zond V	Sep. 15, 1968	Tyuratam	D-1-e	4,820	First lunar flyaround, return, and recovery. Carried biological specimens, photographed Earth. Returned to Earth Sep. 21, 1968; was recovered from Indian Ocean.
Zond VI	Nov. 10, 1968	Tyuratam	D-1-e	4,820	Second unmanned circumlunar flight and recovery. Carried biological specimens, camera. Landed in U.S.S.R. Nov. 17 after double-dip glide reentry, aerodynamic lift for deceleration.

REFERENCES

I. BIBLIOGRAPHICAL REFERENCES

Dickson, Katherine M. *History of Aeronautics and Astronautics: A Preliminary Bibliography, NASA HHR-29.* Springfield, Va.: National Technical Information Service, 1968.

National Aeronautics and Space Administration. *Lunar Surface Studies: A Continuing Bibliography, NASA SP-7003.* Springfield, Va.: National Technical Information Service, January 1962 - March 1964 to date.

II. CHRONOLOGIES

National Aeronautics and Space Administration. *Astronautics and Aeronautics, 1963, NASA SP-4004.* Washington, D.C.: Government Printing Office, 1964.

Astronautics and Aeronautics, 1964, NASA SP-4005. 1965.

Astronautics and Aeronautics, 1965, NASA SP-4006. 1966.

Astronautics and Aeronautics, 1966, NASA SP-4007. 1967.

Astronautics and Aeronautics, 1967, NASA SP-4008. 1968.

III. ARCHIVES

National Aeronautics and Space Administration, Washington, D.C. (Older material is retired to the Federal Records Center, but may be recalled for reference, through the appropriate office.)

NASA History Office Archives.

Office of Manned Space Flight.

    Apollo Program Office File
    Apollo Site Selection Board File.

Office of Space Science and Applications (divided in 1971 into Office of Space Science and Office of Applications).

    Lunar Orbiter Program Office File.
    Planetology Subcommittee File.
    Surveyor/Orbiter Utilization Committee File.
    Surveyor Program Office File.

NASA Langley Research Center, Langley Station, Hampton, Virginia.

Lunar Orbiter Project Office File.

IV. PRIMARY DOCUMENTS

Documents are listed alphabetically under each chapter heading.

CHAPTER I. UNMANNED LUNAR EXPLORATION AND THE NEED FOR A LUNAR ORBITER

Hibbs, Albert R. (ed.). *Exploration of the Moon, the Planets, and Interplanetary Space*, JPL Report No. 30-1. Pasadena, Calif.: Jet Propulsion Laboratory, April 30, 1959.

United States House of Representatives, Committee on Science and Astronautics. *Investigation of Project Ranger, Hearings before the Subcommittee on NASA Oversight*. 88th Congress, 2nd Session, April 27, 29, 30; May 4, 1964.

Report of the Subcomittee on NASA Oversight. June 16, 1964.

Urey, Harold C. "The Chemistry of the Moon," *Proceedings of the Lunar and Planetary Exploration Colloquium*, Vol. I, No. 3. Held at Jet Propulsion Laboratory, October 29, 1958

CHAPTER II. TOWARD A LIGHTWEIGHT LUNAR ORBITER

Boyle, W. S. Bellcomm Working Paper, to J. F. Shea, May 10, 196

Cortright, Edgar M. Memorandum for Messrs. Nicks, Cunningham, Kochendorfer, Mitchell, Subject: Briefing of Seamans on current program proposals, May 15, 1963.

Cummings, Clifford I. Memorandum of Director of Lunar Orbiter Programs, JPL, to Oran W. Nicks, Director, Office of Lunar and Planetary Programs, NASA Headquarters, October 26, 196

Cunningham, Newton W. Memorandum to Charles Sonett, NASA Headquarters, January 12, 1962

Dobies, Edwin F. "The Lunar Orbiter Photographic Experiment," Jet Propulsion Laboratory Section Report No. 1-48. June 1, 1960.

Hughes Aircraft Company. Document No. 262001, June 18, 1962.

Langley Research Center. Project Development Plan for Lunar Orbiter Project (updated December 1964 and June 10, 1966), Project No. 814-00-00.

    Status Report on Lunar Orbiter, August 1, 1963.

NASA Management Manual, Part I, General Management Instructions. March 8, 1963.

Newell, Dr. Homer E., and D. Brainerd Holmes. Memorandum for the Associate Administrator, NASA, October 22, 1962.

    Memorandum, NASA Headquarters, November 1, 1962.

    Memorandum to the Director, Office of Space Flight, concerning questions on unmanned lunar orbiter, March 14, 1963.

    Memorandum to Dr. Floyd L. Thompson, Director of the Langley Research Center, July 1, 1963.

Nicks, Oran W. Memorandum to Capt. Lee R. Scherer, OSS, September 21, 1962.

    Memorandum to Clifford I. Cummings, November 8, 1962.

Office of Manned Space Flight, NASA. Summary of OMSF Data Requirements Document, no date.

    Discussion of Lunar Surface Photographic Requirements, Appendix III. April 19, 1963.

Office of Space Sciences, NASA (later Office of Space Science and Applications).
<u>Surveyor Orbiter Guidelines.</u> BM;Lw, July 20, 1962.

    Project Approval Document drawn up by Lee R. Scherer. October 16, 1962.

    Status Report on Orbiter, February 28, 1963.

    Memorandum to SL Files from SL/Assistant to the Director for Manned Space Flight Support, Subject: Meeting on Incentive Contracting for Lunar Orbiter at Langley

Research Center, June 25, 26, 1963.

Headquarters Comments on Documents for the RFP of the Agena-class Lunar Orbiter, no date.

Memorandum from SD/Deputy Director, OSS, to S/Director, OSS, concerning: Recommended reprogramming within the Office of Space Sciences, April 25, 1963.

Shea, Joseph F. Memorandum to Oran W. Nicks, Office of Space Sciences, October 23, 1962.

Scherer, Lee R. <u>Study of Agena-based Lunar Orbiters</u>. NASA Headquarters, Office of Space Sciences, October 25, 1962.

Memorandum to Oran W. Nicks, OSS, concerning STL Proposal Nos. SC 5100 and SC 5101, November 16, 1962.

Letter to Clinton E. Brown, Langley Research Center, May 24, 1963.

Letter to Oran W. Nicks and Edgar M. Cortright, Office of Space Sciences, August 23, 1963.

Thiel, A. K. Letter to Oran W. Nicks, Director, Lunar and Planetary Programs, OSS/NASA, Washington, D.C., September 20, 1962.

Thompson, Dr. Floyd L. Letter to NASA Headquarters, Code SL, Attention Capt. Lee R. Scherer, March 6, 1963.

Memorandum to Dr. Eugene M. Emme, NASA Historian, NASA Headquarters, Subject: Comments on draft of Lunar Orbiter History, dated November 4, 1969, December 22, 1969.

Transcript of Proceedings -- Discussion between Nicks, Milwitzky, Scherer, Rowsome, and members of the National Academy of Public Administration, NASA Headquarters, September 12, 1968.

CHAPTER III   BEGINNING THE LUNAR ORBITER PROGRAM

Foelsche, Dr. Trutz. "Remarks on Doses Outside the Magnetosphere, and on Effects Especially on Surfaces and Photographic Films," paper presented at the Meeting to Discuss Charged Particle Effects, NASA, Office of Advanced Research and Technology, March 19-20, 1964, Washington, D.C.

Langley Research Center. Agena Class Lunar Orbiter Photographic Project Plan for Evaluation of Offerors' Proposals, Approved: Eugene C. Draley, Chairman, Source Evaluation Board, September 20, 1963.

    Memorandum for Lunar Orbiter Contract File, Subject: Debriefing of the Hughes Aircraft Company, Culver City, California, January 21, 1964.

    Memorandum for Lunar Orbiter Contract File, Subject: Debriefing of the Lockheed Missiles and Space Company, Sunnyvale, California, January 21, 1964.

    Memorandum for Lunar Orbiter Contract File, Subject: Debriefing of the Martin Company, January 21, 1964.

    Memorandum for Lunar Orbiter Contract File, Subject: Debriefing of the Space Technology Laboratories, Inc., Redondo Beach, California, January 22, 1964.

Office of Space Sciences, NASA. Lunar Orbiter Status Report, OSS Review, September 4, 1965.

    Memorandum from the Office of Lunar and Planetary Programs, NASA Headquarters, to Clifford Nelson, Project Director, Lunar Orbiter Office, Langley Research Center, October 22, 1963.

    OSSA Review -- Lunar Orbiter Status Report, January 23, 1964.

Scherer, Lee R. Memorandum to the Record, September 20, 1963.

Seamans, Jr., Dr. Robert C. Memorandum to Dr. Eugene M. Emme, NASA Historian, Washington, D.C. Comments on "Lunar Orbiter: A Preliminary History," Comment Edition (HHN-71); November 25, 1969.

United States House of Representatives, Committee on Appropriations. <u>Independent Offices Appropriations for 1964, Hearings before a Subcommittee of the Committee on Appropriations</u>. 88th Congress, 1st Session, August 19-20, 1963.

    Committee of Conference. <u>NASA Authorization for Fiscal Year 1964, Conference Report</u> (House Report 706). August 26, 1963.

CHAPTER IV. NASA AND BOEING NEGOTIATE A CONTRACT

The Boeing Company. Biographical note on George H. Hage.

Biographical note on Robert J. Helberg.

Biographical note on Carl A. Krafft.

Cortright, Edgar M. Memorandum to Earl D. Hilburn, April 8, 1964.

Hall, J. R. (ed.). TDS Final Report, Tracking and Data System Report Series for Lunar Orbiter Project, Vol. I, Support Summary (608-15). Pasadena, Calif.: Jet Propulsion Laboratory, September 1, 1969.

Hilburn, Earl D. Memorandum to Edgar M. Cortright, Deputy Associate Administrator for Space Science and Applications, March 19, 1964.

James, Lt. Col. Clifton E. Memorandum to the Under Secretary of the Air Force, February 26, 1964.

Lunar Orbiter Program Office, NASA Headquarters. "Plans for Lunar Orbiter Data Acquisition and Analysis." March 20, 1964.

Newell, Dr. Homer E. Memorandum to Dr. Robert Seamans, Associate Administrator of NASA, March 19, 1964.

Northrop Space Laboratories. Technology Utilization Review and Analysis, Final Report, Vol. II, NSL 64-192. September 1964.

Office of Space Science and Applications, NASA. OSSA Review - Lunar Orbiter Status Report, January 23, 1964.

Status of Lunar Orbiter Program for possible use in OSSA Review, February 24, 1964.

Lunar Orbiter Program Status Report, OSSA Review, March 26, 1964.

Lunar Orbiter Status Report, OSSA Review, May 5, 1964.

Rechtin, Dr. Eberhardt, Director, Advanced Research Projects Agency, Washington, D.C. Letter to Dr. Eugene M. Emme, NASA Historian, November 18, 1969, with comments on manuscript "Lunar Orbiter: A Preliminary History" (NASA HHN-71).

Thompson, Dr. Floyd L. Memorandum to NASA Code S, Attention: Homer E. Newell, Subject: Request for Additional Support for Lunar Orbiter from JPL, April 2, 1964.

CHAPTER V. EARLY FUNDING CONSIDERATIONS

Langley Research Center. Minutes of Lunar Orbiter Program Funding Meeting, Langley Research Center, August 19, 1964.

Lunar Orbiter Project Office, Langley Research Center. <u>Project Lunar Orbiter, Narrative Analysis</u>, August 14, September 4, 14, October 28, December 9, 1964; January 25, February 8, 1965.

Newell, Dr. Homer E. Memorandum to Floyd L. Thompson, Subject: Guidelines for Lunar Orbiter Project, October 22, 1964.

Nicks, Oran W. Memorandum to Floyd L. Thompson, Director of the Langley Research Center, August 20, 1964.

Memorandum to the Director of Program Review and Resources Management, August 21, 1964.

Office Of Space Science and Applications, NASA. Memorandum Subject: Lunar Orbiter Funding, POP-64-3, August 24, 1964.

Scherer, Lee R. Memorandum to Oran W. Nicks concerning Lunar Orbiter FY 1966 Funding, September 4, 1964.

Memorandum to Clifford H. Nelson, Lunar Orbiter Project Manager, Langley Research Center, December 31, 1964.

CHAPTER VI THE LUNAR ORBITER SPACECRAFT

The Boeing Company. <u>The Lunar Orbiter</u>, prepared for NASA Langley Research Center by the Space Division of the Boeing Company. Seattle, Wash.: revised April 1966.

Kosofsky, Leon J., and G. Calving Broome. "Lunar Orbiter: A Photographic Satellite," <u>Journal of the SMPTE</u>, Vol. 74 (September 1965), pp. 775-777.

Office of Space Science and Applications, NASA. OSSA Review -- Lunar Orbiter Program Status Report, May 5, 1965.

OSSA Review -- Lunar Orbiter Program Status Report, July 7, 1964.

Summary of First Quarterly Review, August 26-27, 1964.

OSSA Review -- Lunar Orbiter Program Status Report, September 1, 1964.

CHAPTER VII. BUILDING THE SPACECRAFT: PROBLEMS AND RESOLUTIONS

Foelsche, Dr. Trutz. "Radiation Measurements in LO I - V (Period August 10, 1966 - January 30, 1968)," NASA Langley Research Center, paper presented at the Manned Spacecraft Center Seminar, Houston, Texas, June 21, 1968.

Gurtler, Charles A., and Gary W. Grew. "Meteoroid Hazard near Moon," Science, Vol. 161 (August 2, 1968), pp. 462-464.

Helberg, Robert J., and Clifford H. Nelson. "The Lunar Orbiter -- An Integrated Design," a paper presented at the XVIII International Astronautical Congress, Belgrade, Yugoslavia, September 27, 1967.

Lewis Research Center, Cleveland, Ohio. News Release 65-2, January 6, 1965.

Lloyd, Douglas D., and Robert F. Fudali. "Lunar Orbiter Mission Planning," Bellcomm TR-65-211-1. January 25, 1965.

Lunar Orbiter Project Office, Langley Research Center. Project Lunar Orbiter, Narrative Analysis, December 9, 1964; January 25, February 8, March 17, April 16, 28, 1965

Lunar Orbiter Program Office, NASA Headquarters. Memorandum to the Record, Lunar Orbiter Discussion with Dr. Gordon MacDonald, September 24, 1963, October 2, 1963.

Memorandum to the Record, Summary of Lunar Orbiter Trajectory Meeting, Langley Research Center, April 17, 1964.

Report on the LRC and LeRC Lunar Orbiter Shroud and Adapter Meeting, January 5, 1965.

Memorandum from SL/Engineer, Lunar Orbiter Program, Lunar and Planetary Programs, to Langley Research Center, Attention: Mr. Israel Taback, Lunar Orbiter

Project Office, March 4, 1965.

Memorandum from SL/Engineer, Lunar Orbiter Program, to SL/Manager, Lunar Orbiter Program, March 11, 1965.

Newell, Dr. Homer E. Memorandum to Dr. Floyd L. Thompson, Langley Research Center, October 23, 1964.

Office of Space Science and Applications, NASA. Minutes Working Group on Selenodesy, NASA, May 4, 1962.

Summary of First Quarterly Review, August 26-27, 1964.

Third Quarterly Review, February 24-26, reported March 2 1965.

OSSA Review -- March 9, 1965; April 13, 1965.

Rechtin, Dr. Eberhardt, Director, Advanced Research Projects Agency, Washington, D.C. Letter to Dr. Eugene M. Emme, NASA Historian, November 18, 1969, with comments on manuscript "Lunar Orbiter: A Preliminary History" (HHN-71).

Scherer, Lee R. Memorandum to Oran W. Nicks and Edgar M. Cortright, Subject: Immediate Need for JPL Support for Orbiter, July 10, 1964.

Swetnick, Dr. Martin J. Memorandum to the Record, Subject: Summary Minutes, Lunar Orbiter Meeting at NASA Headquarters, June 10, 1964, document dated June 22, 1964.

Report on Trip to Boeing on October 27-29, 1964, November 5, 1964.

"Unmanned Lunar Scientific Missions, a Summary," November 17, 1964.

Thompson, Dr. Floyd L. Memorandum to Dr. Homer E. Newell, Subject: Request for additional support for Lunar Orbiter from Jet Propulsion Laboratory, April 2, 1964.

CHAPTER VIII. LUNAR ORBITER MISSION OBJECTIVES AND APOLLO REQUIREMENTS

Bellcomm, Inc. "Lunar Orbiter Mission Planning," January 25, 1965.

"Apollo Lunar Site Analysis and Selection," March 30, 1965.

The Boeing Company. Boeing Quarterly Technical Progress Report, April to June 1965.

Boeing Quarterly Technical Progress Report, July to September 1965.

Byrne, C. J. Tape Recording of Lunar Orbiter Pictures, Technical Memorandum 65-1012-6. Bellcomm, Inc., July 6, 1965.

Crabill, Norman L., Mission Analysis and Design Engineer, Viking Project Office, Langley Research Center. Memorandum to NASA Code EH, Attention: Dr. Eugene M. Emme, December 9, 1969.

Crabill, Norman L. and A. Thomas Young. "Preliminary Lunar Orbiter Mission Types." Lunar Orbiter Project Office, July 16, 1965.

Howard, Brian T. Memorandum to Maj. General Samuel C. Phillips, NASA/MA, May 10, 1965.

James, Dennis B., Bellcomm, Inc., Memorandum for File, June 30, 1965, Subject: Trip Report: Lunar Orbiter Mission Planning Meeting -- Langley Research Center, June 25, 1965.

Kosofsky, Leon J. Memorandum to Lunar Orbiter Operations Working Group (SL), Subject: Potential Conflict in Goldstone Support of Lunar Orbiter Performance Demonstration Test and Pioneer Mission A, November 22, 1965.

Liddel, Dr. Urner. Memorandum to SL/Director, Lunar and Planetary Programs, Subject: Resolution on Lunar Orbiter Scientific Missions, November 5, 1965.

Lunar Orbiter Program Office, NASA Headquarters. Minutes: Lunar Orbiter Target Objectives Meeting at Langley Research Center, June 25, 1965.

Lunar Orbiter Mission Planning Meeting, Langley Research Center, September 8-9, 1965.

Lunar Orbiter Project Office, Langley Research Center. Lunar Orbiter Project Office Recommendation for Lunar Orbiter Mission A, presented to the Ad Hoc Surveyor/Orbiter Utilization Committee, September 29, 1965.

Project Lunar Orbiter, Narrative Analysis, February 8; March 17, 31; April 22, 28; May 12; June 7, 23; July 2, 30; August 18; November 12, 1965.

National Aeronautics and Space Administration, Washington, D.C. NASA Negotiated Contract No. NAS 1-3800, May 7, 1964.

Newell, Dr. Homer E. Memorandum to AD/Deputy Administrator, with telegram attached, March 9, 1966.

Nicks, Oran W. Memorandum to MA/Maj. General Phillips, Office of Manned Space Flight, September 22, 1964.

Memorandum to MA-6/L. Reiffel, Apollo Program Office, April 26, 1966.

Office of Space Science and Applications, NASA. OSSA Review -- May 6; June 7; July 2, 30; September 9; October 5; November 2, 1965.

SSA/MSF Site Survey Meeting, Minutes, August 4, 1965.

Summary Minutes: Planetology Subcommittee of the Space Science Steering Committee, October 21-22, 1965.

Planetology Subcommittee of the Space Science Steering Committee, Meeting No. 4-66, May 9-11, 1966.

Phillips, Maj. Gen. Samuel C. Memorandum to SL/Lunar and Planetary Programs Director, February 18, 1965.

Rechtin, Dr. Eberhardt, Director, Advanced Research Projects Agency, Washington, D.C. Letter to Dr. Eugene M. Emme, NASA Historian, November 18, 1969, with comments on manuscript "Lunar Orbiter: A Preliminary History" (HHN-71).

Reiffel, Leonard. Memorandum to SL/O. W. Nicks, Subject: Project Apollo Requirements for Lunar Orbiter Data, April 4, 1966.

Scherer, Capt. Lee R. Memorandum to Langley Research Center, Attention: Mr. C. H. Nelson, and Mr. S. L. Butler, October 28, 1965.

Memorandum to SL/Director, Lunar and Planetary Programs, March 7, 1966.

Memorandum to SL/O. W. Nicks concerning update of Orbiter status, April 7, 1966.

Surveyor/Orbiter Utilization Committee. Minutes First Meeting, Washington, D. C., August 20, 1965.

Thompson, T. H. Bellcomm, Inc. Memorandum to Dr. G. E. Mueller/Maj. Gen. S. C. Phillips, December 23, 1964.

    Attachment A -- Review of Current Status of Work Related to Lunar Site Analysis and Selection, December 23, 1964.

Young, A. Thomas. Memorandum to N. L. Crabill, Langley Research Center, May 7, 1965, Subject: Mission Reliability Analyses and Comparison for the Bellcomm Mission and TBC's S-110 Mission.

    Memorandum to N. L. Crabill, Langley Research Center, June 14, 1965, Subject: Lunar Orbiter Mission Planning Study.

CHAPTER IX. MISSIONS I, II, III: APOLLO SITE SEARCH AND VERIFICATION

The Boeing Company. Boeing Quarterly Technical Progress Report, Lunar Orbiter Program, July to September 1966.

    Lunar Orbiter I Final Mission Report, Vol. III, *Mission Operational Performance*, Boeing Document D2-1007-3. February 3, 1967.

    Boeing Quarterly Technical Progress Report, January to March 1967.

Crabill, Norman L., Mission Analysis and Design Engineer, Viking Project Office, Langley Research Center. Memorandum to NASA Code EH, Attention: Dr. Eugene M. Emme, December 9, 1969.

Hall, J. R. (ed.). *TDS Final Report*, Tracking and Data System Report Series for the Lunar Orbiter Project, six volumes. Pasadena, Calif.: Jet Propulsion Laboratory, 1969.

    Vol. I, *Support Summary* (608-15). September 1, 1969.
    Vol. II, *Mission A Summary* (608-17). November 15, 1969.
    Vol. III, *Mission B Summary* (608-18). November 15, 1969.
    Vol. IV, *Mission C Summary* (608-19). November 1, 1969.
    Vol. V, *Mission D Summary* (608-20). November 15, 1969.
    Vol. VI, *Mission E Summary* (608-21). December 15, 1969.

Lunar Orbiter Program Office, NASA Headquarters. Minutes of the Lunar Orbiter Mission B Planning Meeting, Langley Research Center, May 6, 1966.

Lunar Orbiter Mission Status Report 8, August 18, 1966, through Report 20, September 1, 1966.

Lunar Orbiter C Mission Objectives, unsigned memorandum, January 24, 1967.

Status of Lunar Orbiter III, reports February 9, 1967, through March 1, 1967.

Lunar Orbiter Project Office, Langley Research Center.
    Lunar Orbiter Mission B Description. June 1, 1966.

    Lunar Orbiter Mission II Description as amended September 29, 1966. Issued October 26, 1966.

    Lunar Orbiter Mission III Description. January 25, 1967.

    Distribution Chart for Lunar Orbiter Photography. June 15, 1967.

    Project Lunar Orbiter, Narrative Analysis, August 3, 17, 1966; January 17, February 15, 1967.

National Aeronautics and Space Administration. Lunar Orbiter I Photographic Mission Summary, NASA CR-782. Washington, D.C.: April 1967.

    Lunar Orbiter I -- Photography, NASA CR-847. August 1967.

    Lunar Orbiter II Photographic Mission Summary, NASA CR-883. October 1967.

    Lunar Orbiter III Photography, NASA CR-984. February 1968.

Scherer, Capt. Lee R. Memorandum to the File, Subject: Preshipment Review of Second Lunar Orbiter Flight Spacecraft, May 24, 1966.

    Memorandum to File, Subject: Lunar Orbiter I situation, October 28, 1966.

    Memorandum to Clifford Nelson, Langley Research Center, Subject: Geometric Calibration of High Resolution Camera for Mission C, December 20, 1966.

    Memorandum to SE/Deputy Associate Administrator for Space Science and Applications (Engineering), January 24, 1967.

CHAPTER X. MISSIONS IV AND V: THE LUNAR SURFACE EXPLORED

Foelsche, Dr. Trutz. "Radiation Measurements in LO I - V (Period August 10, 1966 - January 30, 1968)," NASA Langley Research Center, paper presented at the Manned Spacecraft Center Seminar, Houston, Texas, June 21, 1968.

Lunar Orbiter Program Office, NASA Headquarters. Minutes of the Mission V Planning Group, NASA Headwuarters, March 7, 1967; and May 26, 1967.

Post Launch Mission Operation Report (MOR) No. S-814-66-04: Lunar Orbiter IV Post Launch Report #1, March 7, 1967, through #13, June 5, 1967.

Lunar Orbiter Mission V Description, approved by the Ad Hoc Surveyor/Orbiter Utilization Committee on June 14, 1967.

Status of Lunar Orbiter E, July 27, 1967.

Post Launch Mission Operation Report No. S-814-67-07: Lunar Orbiter V Post Launch Report #1, August 2, 1967, through #11, September 7, 1967.

Termination of Active Lunar Orbiters: Present Plans for Terminating Active Lunar Orbiters II through V, Lunar Orbiter Item 29, September 11, 1967.

Lunar Orbiter Project Office, Langley Research Center. Lunar Orbiter Mission IV Description, April 26, 1967.

Lunar Orbiter Project Mission Countdown Document LOTD-106-4, July 5, 1967.

Project Lunar Orbiter, Narrative Analysis, March 15, April 17, July 13, 18, 1967.

National Aeronautics and Space Administration. Lunar Orbiter V Photography, NASA CR-1094. Washington, D.C.: June 1968.

Lunar Orbiter V Extended Mission Spacecraft Operations and Subsystem Performance, NASA CR-1142. August 1968.

Executive Secretariat, Program and Special Reports Division, Space Flight Record, 1958-1968. December 31, 1968.

Newell, Dr. Homer E. NASA Mission Objectives for Lunar Orbiter I signed by Edgar M. Cortright for Homer E. Newell, July 25, 1967.

Scherer, Capt. Lee R. Memorandum to SL/Director, Lunar and Planetary Programs, Subject: Lunar Orbiter Mission 5 Planning, March 9, 1967.

Memorandum to SE/Deputy Administrator for Space Science and Applications (Engineering), April 14, 1967.

Memorandum to SL/D. Pinkler, Subject: Lunar Orbiter Program Highlights, September 13, 1967.

Swetnick, Dr. Martin J. Memorandum to the File, Subject: Status of assessment of Lunar Orbiter IV radiation detector data, June 1, 1967.

CHAPTER XI. CONCLUSIONS: LUNAR ORBITER'S CONTRIBUTION TO SPACE EXPLORATION

Apollo Program Office, NASA Headquarters. Minutes of Apollo Site Selection Board, December 15, 1966, document dated March 7, 1967.

Attachment G, Preliminary Landing Site Analysis of Orbiter I.

Minutes of the Apollo Site Selection Board Meeting, March 30, 1967, document dated June 26, 1967.

Minutes of the Apollo Site Selection Board Meeting, December 15, 1967, document dated January 29, 1968.

Foelsche, Dr. Trutz. "Radiation Measurements in LO I - V (Period August 10, 1966 - January 30, 1968)," NASA Langley Research Center, paper presented at the Manned Spacecraft Center Seminar, Houston, Texas, June 21, 1968.

Gault, D. E.; E. M. Shoemaker; and H. J. Moore. *Fragments Ejected from Lunar Surface by Meteoroid Impact Analyzed on Basis of Studies of Hypervelocity Impact in Rock and Sand*, NASA Technical Note TN-D-1767. Washington, D.C.: 1963.

Gurtler, Charles A., and Gary W. Grew. "Meteoroid Hazard near Moon," *Science*, Vol. 161 (August 2, 1968), pp. 462-464.

Hawkins, G. S. *Monthly Notices of the Royal Astronomical Society*, Vol. 116, No. 1 (1956), p. 92.

Lunar Orbiter Program Office, NASA Headquarters. Action Item Summary, Action Item Number 31, Lunar Orbiter: Review and report the necessity for an additional Lunar Orbiter Mission, memo date June 16, 1967.

    Comments on Seamans Draft Memo (undated), June 26, 1967.

Lunar Orbiter Project Office, Langley Research Center. Memorandum to NASA, Code SL, Attention: Capt. L. R. Scherer, Subject: Lunar Orbiter Project Recommendation for Implementing an Additional Mission, July 12, 1967.

National Aeronautics and Space Administration. <u>Lunar Orbiter I -- Photography</u>, NASA CR-847. Washington, D.C.: August 1967.

    <u>Lunar Orbiter V -- Photography</u>, NASA CR-1094. June 1968.

Nazarova, T. N.; A. K. Rybakov; and C. D. Komissarov. "Investigation of solid interplanetary matter in the vicinity of the Moon," paper at 10th COSPAR meeting, London, July 1967.

Newell, Dr. Homer E. Memorandum to AD/Deputy Administrator, Subject: Considerations related to decision on a sixth Lunar Orbiter, July 14, 1967.

Nicks, Oran W. "Applying Surveyor and Lunar Orbiter Techniques to Mars," address before American Institute of Aeronautics and Astronautics, Washington, D.C., December 5, 1968.

Office of Manned Space Flight, NASA Headquarters. Minutes of the Joint Meeting of the Apollo Site Selection Board and the Surveyor/Orbiter Utilization Committee, June 1, 1966, document dated July 1, 1966.

Phillips, Maj. Gen. Samuel C. Memorandum, Subject: Minutes of Apollo Site Selection Board Meeting, March 16, 1966, document dated May 5, 1966.

    Memorandum, Subject: Minutes of the Apollo Site Selection Board Meeting, March 30, 1967.

Scherer, Capt. Lee R. Memorandum to SL/Acting Director, Lunar and Planetary Programs, Subject: Lunar Orbiter 6, April 6, 1967.

    Telegram, priority, unclassified, to Langley Research Center, Attention: Dr. F. L. Thompson, Mr. E. C. Draley, Mr. C. H. Nelson, Jul 24, 1967.

V. INTERVIEWS

The author interviewed the following officials about their roles in the Lunar Orbiter Program.

Bellcomm, Inc., Washington, D.C.

    James, Dennis B., July 25, 1967; July 25, 28, 1969.

The Boeing Company, Washington, D.C.

    Costello, Thomas R., recorded interview, July 9, 1970.

National Aeronautics and Space Administration, Washington, D.C.

    Kosofsky, Leon J., Lunar Orbiter Program engineer, Office of Space Science and Applications, July 5, 1967.

    Liddel, Dr. Urner, Special Assistant, Office of Space Science and Applications, July 14, 1969.

    Newell, Dr. Homer E., NASA Associate Administrator for Space Science and Applications, August 24, 1967.

    Nicks, Oran W., Director of Lunar and Planetary Programs, Office of Space Science and Applications, August 14, 1967.

    Scherer, Capt. Lee R., Lunar Orbiter Program Manager, Office of Space Science and Applications, July 31, 1967 (en route to Cape Kennedy, Florida, from Orlando, Florida, for *Lunar Orbiter V* launch); August 14, 1967.

    Thompson, Dr. Floyd L., former Director of Langley Research Center, January 29, 1970.

Lunar Orbiter Project Office, Langley Research Center, Langley Station, Hampton, Virginia.

    Brewer, Gerald W., Lunar Orbiter Mission Assurance, July 18, 1967; July 7, 1970.

    Broome, G. Calvin, Photo Subsystem Manager, July 19, 1967.

Elder, Theodore H., Technical Administration, July 18, 1967.

Foelsche, Dr. Trutz, Aeronautical and Space Mechanics Division, July 7, 1970.

Girouard, Robert L., Space Vehicle System, July 18, 1967.

Graham, John B., Operations Integration, July 19, 1967.

Katzoff, Dr. Samuel, Chairman of Lunar Orbiter Advisory Group, telephone interview, August 24, 1967.

Martin, James S., Jr., former Assistant Lunar Orbiter Project Manager (later Project Manager, Viking Project), recorded interview, July 7, 1970.

Nelson, Clifford H., Lunar Orbiter Project Manager, July 20, 1967.

Recant, Isadore G., Data Handling, July 20, 1967.

Taback, Israel, Lunar Orbiter Spacecraft Manager, July 7, 1970.

Young, A. Thomas, Mission Integration, July 21, 1967.

VI. ADDITIONAL READING

This section consists of annotated references selected by the author to give a cross-section of information on the Lunar Orbiter spacecraft, its mission, and lunar scientific exploration from 1961 to 1969. Many more articles and publications about these subjects exist. However, the author has selected these because most of them pertain to data acquired from the five Lunar Orbiter missions. The list is intended to give the reader a general survey of hypotheses, theories, and arguments about the origins, the nature and the surface features of the Moon which Lunar Orbiter has helped to uncover. It is hoped that this will arouse the reader's curiosity to investigate the realm of lunar sciences and exploration further.

Adler, J. E. M., and J. W. Salisbury. "Behavior of Water in Vacuum: Implications for 'Lunar Rivers,'" Science, Vol. 164 (May 2, 1969), p. 589.

>The investigators conducted laboratory experiments using soils with grain sizes ranging from 0 to 125 microns and gravels ranging from 2 to 4 millimeters with gradations and layering. Tests were run under air and vacuum conditions to determine behavior of water at various flow rates and temperature levels on test soils. Results showed that, in the presence of air, water formed terrestrial-like stream channels. In a vacuum at freezing temperatures water formed dendritic ice masses and continued to flow under the ice, frequently penetrating to the surface and freezing. Water then sublimated, leaving a hummocky surface. Some soil downslope movement occurred, but no stream channels developed. Results show that ice will readily form in a vacuum to a thickness which allows liquid water to exist under it. Model streams produced in a vacuum did not erode rille-like channels. Results support Lingenfelter's predictions (Science, Vol. 161, p. 266).

Alfvén, H. "Origin of the Moon," Science, Vol. 148 (April 23, 1965), pp. 476-477.

>There is a major implication in the mathematical calculations of the Moon's orbit as rechecked and improved by H. Gerstenkorn. About one billion years ago the Moon, a separate planet orbiting the Sun, passed very close to Earth. Both bodies continued to attract each other until the Moon assumed a retrograde orbit about the rapidly spinning Earth. The Moon moved within the Roche limit in a polar orbit around

Earth, causing part of the lunar surface to break away and bombard Earth. Following this the Moon began to recede from Earth until it came to occupy its present orbit. Loosened materials fell back on the Moon as meteors, making major craters. Geological investigations might substantiate Gerstenkorn's theory.

Allen, D. A., and E. P. Ney. "Lunar Thermal Anomalies: Infrared Observations," Science, Vol. 164 (April 25, 1969).

Infrared observations of the Moon in the 8- to 14-micron atmospheric window have delineated macroscopic lunar surface thermal behavior. Shorthill has discovered further lunar thermal anomalies. The craters Aristarchus, Copernicus, and Tycho cool much less rapidly than surrounding areas during eclipse. The observations made by the authors have not determined the geometric scale of the structure of hot and cold regions. Surface rocks in these areas may be responsible for the less rapid cooling rates because they are probably thermally connected to a subsurface temperature of 200 degrees Kelvin.

Bailey, Norman G. Cinder Lake Crater Field Location Test. United States Geological Survey Interagency Report: Astrogeology 2, November 1967.

This report describes the use of Lunar Orbiter II photographs in conducting a test in which the subjects were required to fix the location of a Lunar Module in a simulated crater field near U. S. route 89, northeast of Flagstaff, Arizona.

Baldwin, Ralph B. "Lunar Mascons: Another Interpretation," Science, Vol. 162 (December 20, 1968), pp. 1407-1408.

The author questions the survivability of an impacting body. He postulates that 1) craters formed by impacting events are dry, not lava-filled, 2) isostatic distortions occurred, but before this was complete, lava appeared from the body of the Moon and selectively filled the lower areas. This lava was denser than surrounding rock, which presumably could have been more acidic, and 3) tension cracks (rilles) and compression fractures (wrinkle ridges) show that later subsidence and compression has occurred. Thus far only the dense material centered in craters and capable of yielding gravitational effects has been measured.

The Boeing Company. Final Report on A Study of the Lunar Orbiter Regarding Its Adaptability to Surface Experiments Utilizing a Fly-by and Earth-Return Trajectory. October 6, 1966, prepared for NASA Langley Research Center.

This report outlines the necessary requirements and constraints which would have to be met in order to put a Lunar Orbiter in an Earth-return trajectory around the far side of the Moon. This constitutes the basis of a contingency plan, should the Orbiter have failed to go into orbit around the Moon. During the fly-by the Orbiter could have taken useful photographs of the far side of the Moon. Upon return to the Earth the spacecraft would burn its remaining propellent to deboost into Earth orbit for readout of the data.

Cambell, Malcolm J.; Brian T. O'Leary; and Carl Sagan. "Moon: Two New Mascon Basins," Science, Vol. 164 (June 13, 1969), pp. 1273-1275.

In studying existing spherical harmonic expansions of the Moon's gravitational potential and the difference among the lunar principal moments of inertia, the authors found two large gravitational anomalies not associated with those of Muller and Sjogren. One on the east limb of the Moon near Mare Marginis appears to be associated with a large circular basin, 900 kilometers in diameter, centered at 91 degrees east, 25 degrees north, with Mare Marginis filling in the southwest corner.
On the far side, Lunar Orbiter photos disclose that the authors feel is an enormous circular basin now very heavily eroded. The basin is 1,000 kilometers in diameter, centered at 173 degrees east, 11 degrees north. They propose that this be called Occultum (Hidden Basin).

Cameron, Winifred S. "An Interpretation of Schröter's Valley and Other Lunar Sinuous Rilles," Journal of Geophysical Research, Vol. 69 (June 15, 1964), pp. 2423-2430.

Various theories exist about the origin of lunar sinuous rilles such as Schröter's Valley. The mechanism producing them can be categorized under aqueous erosion, faulting, and subsidence. Each of these does not stand the intensive investigations of the rilles' topography. Aqueous erosion is the least tenable of all the mechanisms because it necessitates the presence of very high vapor pressures for any liquid at lunar surface temperatures. Observable evidence speaks against faulting as the major mechanism causing rilles.

Igneous processes suggest another mechanism, but outflow of lava creates a raised feature, not a depression. Yet one process could explain their formation: nuees ardentes, or fluidized outflows of gas-dust mixtures. The presence of sinuous rilles in the vicinity of craters whose formation seems to be volcanic strongly suggests a relationship supporting this mechanism as the process by which these surface features have been formed.

Cameron, Winifred S.; Paul D. Lowman, Jr.; and John A. O'Keefe. "Lunar Ring Dikes from Lunar Orbiter I," Science, Vol. 155 (January 6, 1967), pp. 77-79.

    Lunar Orbiter I photographs reveal portions of the Flamsteed Ring near the Surveyor I site. The convex body resembling a flow of viscous lava located near Apollo landing site A 9.2 at 2 degrees south latitude, 43 degrees west longitude has partially invaded nine craters in the area. This suggests that the flow material is younger than the mare material. The investigators conclude that these topographic features indicate the presence of extruded intermediate lavas of acidic composition. Such lavas are more viscous than basic lavas. The investigators further conclude that the Flamsteed Ring is not the result of basaltic flows despite lower gravity on the Moon. These conclusions are preliminary.

Conel, J. E., and G. B. Holstrom. "Lunar Mascons: A Near-Surface Interpretation," Science, Vol. 162 (December 20, 1968), pp. 1403-1404.

    The work of these two men shows that near-surface slab-like models produce anomalies of the magnitude observed from tracking data of the Lunar Orbiters. The authors assume that maria fill can be represented by a thin circular disk of dense rock at the lunar surface, imbedded in less dense material. Submare and adjacent rim material has either lower density because this has been breciated and pulverized by impact, or is a high-density material if brought to the impact site by an impacting body.

Elston, Donald P. Character and Geologic Habitat of Potential Deposits of Water, Carbon, and Rare Gases on the Moon. United States Geological Survey Interagency Report: Astrogeology 6, May 1968.

This report concerns geological characteristics of the Moon, general composition, lunar geological processes, and cratering by possible cometary materials. Lunar Orbiter V photographs are used in the analysis of the craters Messier and Messier A.

Elston, Donald P., and Charles R. Willingham. Five-day Mission Plan to Investigate the Geology of the Marius Hills Region of the Moon. United States Geological Survey Interagency Report: Astrogeology 14, April 1969.

Lunar Orbiter V photographs H-216 and H-217 of the Marius Hills constitute the basis for a geological survey which a manned roving vehicle could conduct during a five-day period on the lunar surface. Included in this report are two large geological maps with scales of 1:200,000 and 1:25,000 respectively.

Fielder, G., and J. E. Guest. "Lunar Ring Structures and the Nature of the Maria," Planetary Space Science, Vol. 16 (May 1968), pp. 665-673.

A new interpretation of lunar ring structures is the result of analysis of data from Lunar Orbiter and Surveyor. Instead of accepting the hypothesis that "elementary" rings represent old, partially filled craters, the authors posit the hypothesis that they are recent volcanic structures. Elementary ring structures occur mostly on flat, smooth floors of maria. They consist of lunaritic materials in hills or wrinkle ridges of both. The rings approximate circles or polygons and parts of them coincide in direction with local tectonic patterns. The rings are generally incomplete. The authors do not claim that all incomplete rings on the Moon have the same origins or are of the same type.

Filice, Alan L. "Lunar Surface Strength Estimate from Orbiter II Photograph," Science, Vol. 156 (June 16, 1967), pp. 1486-1487.

A Lunar Orbiter II photograph of an area in western Mare Tranquillitatis shows a boulder track down the wall of the crater Sabine D. Assuming a spherical boulder of $r = 6.5$ meters and a density of 3.0 grams/centimeter^3, then the surface bearing strength is equal to 4 times 10^6 dyne/centimeter^2 at a depth of 75 centimeters. This preliminary measurement is significant because it can be used as a lower limit of bearing strength over a length of 650 meters versus the footpad-sized measurement of a landed spacecraft. The area of this measurement is also significant because it is a potential landing site for Apollo.

Firsoff, V. Axel. "Water Within and Upon the Moon," *New Scientist*, Vol. 37 (March 7, 1968), pp. 528-530.

    Firsoff discusses the implications of Lunar Orbiter photography in relation to two main theories about the formation of lunar surface features: water and volcanic/meteoric. The existence of sinuous rilles, of long valleys and evidences of "aprons" to the west and southwest of Tsiolkovsky suggest water action in various forms from high-pressure sublimation to ash-covered glaciers. Many formations could not have resulted from lava flows as understood by known behavioral characteristics of such flows on Earth. Under conditions on the Moon lava cannot travel far. Water, however, when escaping to the surface under extreme pressure from within, could cause explosions and craters to form. Moreover, if one assumes that Orientale was formed in an astroidal impact event, then this would have released sufficient gases and water trapped within to have formed a temporary lunar atmosphere. The impact would have triggered far-reaching processes and initiated prolonged volcanic activity whose effects would have affected the entire lunar surface.

Fulmer, Charles V., and Wayne A. Roberts. "Surface Lineaments Displayed on Lunar Orbiter Pictures," *Icarus*, Vol. 7 (November 1967), pp. 394-406.

    Lunar Orbiter photography reveals closely spaced parallel lineament sets in such areas as the craters Gambart, Maskelyne F, Gambart C, Kepler, and Copernicus, and also in Oceanus Procellarum and in Marius. These may be surface expressions of underlying faults or fractures. It is not certain if these lineament sets were restricted in formation to a single time span. Lineament sets parallel to polygonal sides or rayed and unrayed craters suggest the presence of a precrater parallel joint system. These surface lineaments may have been produced by Earth tidal forces. This would indicate that the Moon's surface is and has been a working unit through much of lunar history.

Gambell, Neil, and Baerbel K. Lucchitta. *A Limitation of First Generation Lunar Orbiter Negatives as Applied to Photoclinometry*. United States Geological Survey Interagency Report: Astrogeology 11, November 1968.

This report describes tests conducted to determine the usefulness of Lunar Orbiter photographic negatives in determining slopes on the Moon's surface. Random tests were conducted to define the reliability of film density measured against the gray scale. Results show that negatives with density readings higher than step nine of the gray scale give erroneous slope measurements.

Gilvarry, J. J. "Nature of the Lunar Mascons," *Nature*, Vol. 221 (February 22, 1969), pp. 732-736.

Gilvarry posits the theory that positive and negative mascons have been caused by a series of events after the initial formation of the Moon: The lunar seas constitute the oldest exposed areas of the surface. Their presence and the existence of positive and negative gravitational anomalies in irregular maria rule out the lava mechanism formation theory and support the theory of a lunar hydrosphere at some time after the Moon's formation. Experiments with various soil types under conditions involving simulated lunar hydrosphere, atmosphere, and vacuum conditions offer explanations for the nature of maria materials, the former existence of surface water acting as a transport mechanism for these materials, and the differing isostatic conditions between maria and highland areas. Negative mascons would have resulted when overlying water flowed to lower areas or escaped into space. The geographical location of negative mascons supports this supposition. Water, in turn, carried deposits down to the great circular maria whose depths, produced by meteoric impacts, accepted greater sedimentation and, therefore, increased mass concentrations.

Guest, J. E., and J. B. Murray. "Nature and Origin of Tsiolkovsky Crater, Lunar Farside," *Planetary Space Science*, Vol. 17, pp. 121-141. Oxford: Pergammon Press, 1969.

The authors discuss the formation of the Tsiolkovsky crater on the farside of the Moon. They base their observations on data from *Lunar Orbiter III* high- and medium-resolution frames No. 121. Tsiolkovsky is a landmark on the far side, a young, distinct, and very large crater in an area saturated with craters. The authors discuss the probable origins of Tsiolkovsky in relation to: 1) the distribution of craters around it, 2) the nature and shape of its rim, 3) radial gouges and crater chains, and 4) the presence of an apparent ejecta blanket. They conclude that Tsiolkovsky formed as a result of an impacting astroidal body or a giant volcanic explosion, and they prefer the former hypothesis to the latter.

Gurtler, Charles A., and Gary W. Grew. "Micrometeoroid Hazard near Moon," Science, Vol. 161 (August 2, 1968), pp. 462-464.

    All five Lunar Orbiters flew micrometeoroid flux experiments to test the frequency of micrometeoroid hits in the lunar environment. The only other spacecraft which had attempted to do this was the Soviet Luna 10. This spacecraft had registered particle impacts exceeding by two orders of magnitude the average of interplanetary space. The Lunar Orbiter experiments had a configuration which detracted from maximum exposure to the lunar environment. Test material on board each spacecraft consisted of pressurized beryllium copper detectors covering an area of 0.282 square meters, of which only 0.186 square meters was effectively exposed. Over a one year period five Orbiters recorded a total of 22 hits or one-half the record registered in Earth orbit by Explorers 16 and 23, using the same kind of detectors. The investigators caution that these data are too tentative to form a general theory about micrometeoroid flux near the Moon.

Hartmann, W. K. "Lunar Basins, Lunar Lineaments, and the Moon's Far Side," Sky and Telescope, Vol. 32 (September 1966), pp. 128-131.

    Hartmann has examined rectified pictures from the Russian Zond III of portions of the Moon's far side and of Orientale Basin. He discusses the significance of the pictures in theories concerning the formation of lunar basins and the maria. Of special interest is Orientale which involves a whole system of craters, crater chains, concentric mountain rings and scarps including the Rook and Cordillera mountains. Photographic data is still too scarce to determine what role, if any, volcanism, tectonic activity, and ejected rubble played in modifying ancient continental uplands.

Hixon, S. B. "Topography and Geologic Aspects of a Far-Side Lunar Crater," Science, Vol. 159 (January 26, 1968), pp. 420-421.

    This brief article describes a flow-like surface feature in a farside crater some 70 kilometers south of Tsiolkovsky. Initial analysis of Lunar Orbiter photography indicates that the flow has a thickness of at least 20 meters at a point about 4 kilometers east of G in the superimposed schematic on the photograph. The author rules out the possibilities of it being a mudflow or an air-cushioned landslide because of vacuum conditions. He suggests that it is considerably more like an ash-flow tuff.

Hughes, J. Kenrick, and David E. Bowker. Lunar Orbiter Photographic Atlas of the Moon. National Aeronautics and Space Administration, NASA SP-206, 1971.

    A selection of photographs giving complete coverage of the Moon, front and back, and referenced to the surface by index map

Hunt, Graham R.; John W. Salisbury; and Robert K. Vincent. "Lunar Eclipse Infrared Images and an Anomaly of Possible Internal Origin," Science, Vol. 162 (October 11, 1968), p. 252.

    The authors conducted infrared studies of the Moon in eclipse on April 13, 1968, and their observations were the first to confirm the thermal anomalies observed by Saari and Shorthill in December 1964. They conclude that because the hundreds of anomalies have remained unchanged in 3.5 years, they are not the result of ephemeral activity on the lunar surface. They detected a linear thermal anomaly at the western edge of Mare Humorum which, unlike prominent crater anomalies, is warmer than its surroundings before sunset. It remains warmer after sunset. Lunar Orbiter IV photography of Mare Humorum, at a ground resolution of 54 meters, shows no unusual surface structures which would support the belief that the anomaly is caused by low-thermal-inertia material. The more probable cause is an internal heat source because 1) heat flow to the surface would make an area warmer than its surroundings during lunar afternoon, and 2) the geological position of the anomaly supports this.

Karlstrom, T. N. V.; J. F. McCauley; and G. A. Swann. Preliminary Lunar Exploration Plan of the Marius Hills Region of the Moon. United States Geological Survey Interagency Report: Astrogeology 5, February 1968.

    The scientific objectives, operational guidelines and surface exploration constraints of a five-day mission of the Marius Hills constitute the subject of this report. Lunar Orbiter V photographs of this region have been used in constructing preliminary geological maps and descriptions of the traverses which astronauts could perform in a lunar roving vehicle.

Kosofsky, Leon J. "Topography from Lunar Orbiter Photos," Photogrammetric Engineering, Vol. XXXII, No. 2 (March 1966), p. 277.

    The author discusses in detail the Lunar Orbiter photographic mission. Among its major tasks the Orbiter spacecraft is designed to obtain useful topographical data of the lunar surface for the Apollo Program. Special methods of photometric data reduction must be applied to Lunar Orbiter photography because of the peculiar characteristics of reflectivity of the lunar surface. Preflight calibrations will be necessary to compensate for any distortions in high-resolution photography due to the Moon's surface characteristics and the fact that the film will not be returned to Earth.

Kosofsky, Leon J., and Farouk El-Baz. The Moon as Viewed by Lunar Orbiter. National Aeronautics and Space Administration, NASA SP-200, 1970.

> A selected compilation of photographs that illustrate the heterogeneous nature of the lunar surface, including four stereographic views in color and accompanied by index maps. Many features are similar to features on Earth; others have no Earth counterpart. Also included are photographic guideposts for planning manned exploration of the surface.

Lamar, D. L., and Jeannine McGann. "Shape and Internal Structure of the Moon," Icarus, Vol. 5 (1966), pp. 10-23.

> The authors offer a summary of the various theories on the origins of the Moon and its shape and internal composition. They point out that no theory has explained the nature of the Moon's core nor the distribution of the density of subsurface material. They do not suggest the presence of mass concentrations (Mascons) on the Moon.

Lamar, Donald L., and Jeannine V. McGann-Lamar. "Shape and Internal Structure of the Moon, from Lunar Orbiter Data." Earth Science Research Corp., Final Report, NASA Contract NSR 05-264-002, November 1968.

> The report points out that there is a difference between the Moon's center of figure or volume and the center of its mass. There appears to be a systematic excess of elevation of continental areas over maria, relative to the Moon's center of mass. A comparison of the mascons with the lunar map indicates excess masses are concentrated within the inner rings of the Imbrium and Nectare Basins. If mascons are assumed to be masses of nickel-iron, then they correspond to a layer about 12 kilometers thick. Isostatic models of the Moon also fit the data, but Lunar Orbiter data does not sufficiently resolve which model.

Liebelt, Paul B. "The Flight Path Control Software System of the Lunar Orbiter," a paper presented at the International Astronautical Federation, Seventeenth International Astronautical Congress, Madrid, Spain, October 9-15, 1966.

> Ranger and Mariner software programs were found to be inadequate for Lunar Orbiter. Thus the Lunar Orbiter Program developed new concepts for flight control and the necessary software to implement them. Among other things the optimization of the midcourse aim point and the orbit injection point became a necessary and practical procedure. A mean element trajectory program was developed to facilitate orbital transfers by greatly reducing computation times to a few minutes rather than hours as was necessary under the special perturbation analysis approach.

Lingenfelter, Richard E.; Stanton J. Peale; Gerald Schubert. "Distribution of Sinuous Rilles and Water on the Moon," *Nature*, Vol. 220 (December 21, 1968), pp. 1222-1225.

    The authors present a defense of the theory of water on the Moon as the major cause of sinuous rilles. Their analysis is based upon data from *Lunar Orbiter IV* photography and upon Urey's hypothesis of a lunar atmosphere existing at one time in the past. They point out that volcanic ash flows, as suggested by Gold, cannot explain the length and meandering of many rilles. Nor can faulting. However, water flow under a layer of surface ice offers a viable explanation. Moreover, certain events could have caused outgassing of major volatiles H_2O and CO_2. Major meteor impacts would have released trapped volatiles and could have led to a temporary atmosphere. They conclude that the distribution of sinuous rilles is the only available, unambiguous indicator of location of subsurface volatiles.

Lingenfelter, R. E.; S. J. Peale; and G. Schubert. "Lunar Rivers," *Science*, Vol. 161 (July 19, 1968), pp. 266-269.

    Lunar Orbiter photographs show sinuous rilles resembling meandrous channels of terrestrial streams. Thirty of these are visible from Earth. Lunar Orbiter revealed significant new features in the smaller meandrous channels inside the larger rilles. The authors hypothesize that the rilles are features caused by water erosion in the form of ice-covered rivers whose source is subsurface water released through the impacts of meteors.

Lipskii, I. N. "Zond 3 Photographs of the Moon's Farside," *Sky and Telescope*, Vol. 30 (December 1965), pp. 338-341.

    The author describes the achievements of *Luna III* in 1959 and compares them with those of the *Zond III* mission in 1965. The latter confirms the data of the former concerning the lunar far side: it is more heavily cratered than the front side. On the whole the craters exhibit similar features to those on the front side. Crater chains also exist on the far side but are much longer, in some cases 1,500 kilometers. Numerous ring-shaped concavities called thalassoids also can be seen in *Zond III* pictures. In size and shape they compare to maria. No such thalassoids are present on the front side. Lipskii concludes that available data show the Moon's surface to be continental with maria resulting from endogenic depressions being filled with lava.

MacDonald, Gordon J. F. "Interior of the Moon," *Science*, Vol. 133 (April 7, 1961), pp. 1045-1050.

MacDonald discusses the several modern theories concerning the nature and composition of the Moon's interior. He states that even a chemically homogeneous Moon would undergo discontinuities in the structure of subsurface material. Surface features and the lack of evidence of major faulting imply a constant volume of the Moon. Little conclusive evidence exists to prove or disprove current hypotheses. The author suggested a lunar orbiter spacecraft circling the Moon could be tracked and that this would provide data on the Moon's gravitational field, its mean moment of inertia, and other fundamental data which would reveal more about the nature of the Earth's natural satellite.

Mayo, Alton P. "Orbit Determination for Lunar Orbiter," *Journal of Spacecraft and Rockets*, Vol. 5 (April 1968), p. 395.

This report covers the results of orbit determination programs in the first four Orbiter missions. Orbit determination proved to be very accurate and precise with tolerable deviations from planned parameters. Some deviations between planned and executed midcourse, deboost, and orbit maneuvers resulted from oscillation in Doppler residuals, especially in low photographic orbits. Uncertainty of lunar gravitational constraints make orbital statistics not entirely valid. One accomplishment of the program was the improvement of orbit determination as a result of predicted photo-location by real-time and postflight orbit determination. On the *Lunar Orbiter III* mission the difference between the two factors was about 5 kilometers and considerably worse for certain sites in the first two missions.

McCauley, John F. "Geologic Results From the Lunar Precursor Probes," a paper presented at the Fourth Annual Meeting of the American Institute of Aeronautics and Astronautics, October 1967. AIAA Paper No. 67-862.

The author points out that the Lunar Orbiter Program was by far the most productive of the precursor probes in terms of total amount of information received and the nature of that information in certain areas vital to further exploration. The author discusses several of the most significant topographical features which Lunar Orbiter photographed and concludes that the photographic data greatly help in identifying morphological classes of these features.

Michael, William H., Jr., and Robert H. Tolson. "The Lunar Orbiter Project Selenodesy Experiment," a paper presented at the Second International Symposium on The Use of Artificial Satellites for Geodesy, Athens, Greece, April 27-May 1, 1965. NASA/Langley Research Center.

    The authors summarize the mission of Lunar Orbiter and concentrate upon its usefulness in the more refined determination of the lunar gravitational field and the Moon's shape and mass. They briefly review the existing knowledge on these subjects and then describe in detail various technical approaches to the problem of determining spacecraft orbital parameters and what they will show about the Moon.

Michael, William H.; Robert H. Tolson; and John P. Gapcynski. "Lunar Orbiter: Tracking Data Indicate Properties of the Moon's Gravitational Field," Science, Vol. 153 (September 2, 1966), pp. 1102-1103.

    The authors have drawn preliminary conclusions about the significance of the orbital behavior of Lunar Orbiter I based upon early tracking data. Their primary task was the establishment of a rough estimate about the Moon's gravitational field from more extensive data from the other four Lunar Orbiter missions. Preliminary results of their investigation show that orbital variations during periods of photography did not degrade the quality of photographs. Tracking data used in this analysis were two-way Doppler data providing a measure of relative velocity of the spacecraft and the NASA Deep Space Network stations in California, Spain, and Australia.

Mulholland, J. Derral, and William L. Sjogren. "Lunar Orbiter Ranging Data: Initial Results," Science, Vol. 155 (December 9, 1966), p. 74.

    The investigators have used ranging residuals data from the first two Orbiter missions to test corrections in the lunar ephemeris. Most residuals were reduced to less than 100 meters. Preliminary ephemeris tapes at the Jet Propulsion Laboratory were used to analyze raw data. Tracking data from the Deep Space Network stations enabled the investigators to refine the mathematical calculations. Variations in ranging residuals from the three stations verify unusual Doppler residuals obtained near pericenter passage of Lunar Orbiter I. These were not attributed to onboard system anomalies and appeared to be real and to show that the spacecraft had an anomalous motion of 60 meters near pericenter.

Muller, Paul M., and William L. Sjogren. <u>Consistency of Lunar Orbiter Residuals with Trajectory and Local Gravity Effects</u>. JPL Technical Report 32-1307, September 1, 1968.

    The authors have analyzed the results of Earth-based coherent two-way radio Doppler data from the Lunar Orbiters. They found the residuals' consistency to be too high. This could be caused by: 1) forces such as gravity, solar pressure, gas jets; 2) errors in tracking data; and 3) software problems in the computer. They then utilized higher harmonics models of the Moon, and the residuals reduced, reaching agreement between separated flight on the same trajectory.

Muller, Paul M., and William L. Sjogren. "Mascons: Lunar Mass Concentrations," <u>Science</u>, Vol. 161 (August 16, 1968), pp. 680-684.

    The authors have constructed a gravipotential map of the near side of the Moon based upon orbital accelerations of the Lunar Orbiter spacecraft. These show gravitational anomalies termed "mascons" beneath the lunar surface in all five of the ringed maria. This suggests a correlation between mass anomalies and the ringed maria. Conclusions are tentative.

National Aeronautics and Space Administration. <u>Lunar Orbiter I Preliminary Results</u>. NASA report SP-197, 1969.

    A brief description of the Lunar Orbiter Program's history, this report describes the spacecraft, its mission, and what the first Lunar Orbiter accomplished.

Norman, Paul E. "Out-of-This-World Photogrammetry," <u>Photogrammetric Engineering</u>, Vol. XXXV, No. 7 (July 1969), pp. 693-700.

    Norman discusses the Apollo requirements for cartographic and topographic data on the lunar surface, the landing sites, and their approaches. Photogrammetry plays a mandatory role in determining accurate coordinates for landing sites and reference marks called <u>landing-site landmarks</u>. Lunar Orbiter photographic data has provided the only applicable source for making large-scale maps of the Apollo landing zone. How this is done constitutes the subject of the article. The author concludes that Lunar Orbiter successfully demonstrated the potential of surveying and mapping the Moon or a planet from space.

Oberbeck, Verne R., and William L. Quaide. "Estimated Thickness of a Fragmental Surface Layer of Oceanus Procellarum," Journal of Geophysical Research, Vol. 72 (September 15, 1967), p. 469.

    Analyses of Lunar Orbiter I photographs of Oceanus Procellarum showing craters of varying morphology indicate a correlation between crater size and crater shape as a result of meteorite impact against a surface consisting of fragmental material of varying thicknesses overlying cohesive substrata. The analysis of these data indicate that 85% of the area considered has surface thickness between 5 and 15 meters. Photographs from Luna 9 and Surveyor I support this indication. Moreover, formation of new rock surfaces appears to have occurred intermittently, leading to a complex stratigraphic sequence of alternating hard and fragmented rock. The existence of concentric craters substantiates this sequence.

Oberbeck, Verne R., and William L. Quaide. "Genetic Implications of Lunar Regolith Thickness Variations," Icarus, Vol. 9 (1968), pp. 446-465.

    The distribution of the lunar regolith thickness for twelve areas on the Moon has been determined using high-resolution photographs from Lunar Orbiter II, III, and V. All but one area lie within ten degrees of the equator. The exception is in Mare Imbrium. The article compares lunar crater geometry with laboratory craters. Results show that the regolith thickness varied from 3.3 meters in the southern portion of Oceanus Procellarum to 16 meters in the crater Hipparchus. The report also discusses the delineation of flow fronts and the discovery of many linear markings on the presumed flows. These lineaments may be crater chains of a collapsed or drainage origin. Still other lineaments may be lava channels. The authors conclude that the thickness of the regolith is a function of crater density. Over time impacting bodies break down the lunar surface and create the regolith which is the result of impact fragmentation.

Pohn, H. A., and T. W. Offield. Lunar Crater Morphology and Relative Age Determination of Lunar Geological Units. United States Geological Survey Interagency Report: Astrogeology 13, January 1969.

    This report describes a system for determining the relative age of craters on the lunar surface by using as a basis their major topographical components. From this the authors have constructed a preliminary morphological continuum which they use to classify craters over the entire surface of the Moon. Lunar Orbiter photography was instrumental in providing them with reliable data.

Rindfleisch, Thomas. "Photometric Method for Lunar Orbiter," *Photogrammetric Engineering*, Vol. XXXII (March 1966), p. 262.

The photometric method for deriving surface elevations from a single picture of the lunar surface in the absence of stereoscopic pictures is described. The author uses Ranger photographs as subjects and concludes that a derivation of quantitative topographic information about an object scene is possible. At best the resulting data are indirect, and estimation of errors seems unrealistic by analytical means. Moreover, calculations show that it is wrong to assume uniform albedo for large areas.

Rozema, Wesley. *The Use of Spectral Analysis in Describing Lunar Surface Roughness*. United States Geological Survey Interagency Report: Astrogeology 12, December 1968.

Photography from *Lunar Orbiter II*, a topographic map of the *Surveyor III* landing site, and photographs from *Rangers VIII* and *IX* are utilized in applications of the power spectral density (PSD) function to determine relative roughness of different types of lunar terrain. Such information would be valuable in the construction and operation of a lunar roving vehicle.

Scherer, Lee R. "The First Four Lunar Orbiter Photographic Missions," a paper presented to the Committee on Space Research, London, England, July 1967.

Scherer describes the Lunar Orbiter spacecraft as a platform designed to carry a camera system which can take high- and medium-resolution photographs of the Moon's surface. The spacecraft has four objectives: 1) obtain photography of wide areas of the Moon to certify Apollo and Surveyor landing sites, 2) define gravitational field of the Moon through refined tracking of the spacecraft, 3) measure micrometeoroid and radiation flux during extended lifetime of spacecraft, and 4) provide a spacecraft for equipment checkout and personnel training of the Apollo tracking network.

Stipe, J. Gordon. "Iron Meteorites as Mascons," Science, Vol. 162 (December 20, 1968), pp. 1402-1403.

    The author bases his interpretation on studies of impacts of steel projectiles into concrete and soils and then makes large extrapolations upward in size. On the Moon an impacting body must penetrate below the surface to a depth of 290 kilometers before pressure can be released sufficient to melt material. His results suggest that lava-filled maria formed when large iron objects struck the lunar surface at a velocity so low that there was no immediate fracture of the object. The impact produced a large crater and material flowed to the surface to fill the crater. Each mare was formed by one large iron object impacting, and the remnants of this dense object under the mare are the mascon.

Swann, G. A. Lunar Geological Field Investigations. United States Geological Survey Interagency Report: Astrogeology 9, August 1968.

    Swann describes how investigation of the Moon's surface can test the hypotheses based upon terrestrial observations of the geological history of the Earth in an effort to determine the origins of both bodies. The Apollo system constitutes the basic capability with which such extended lunar exploration can be carried out.

Trask, N. J., and L. C. Rowan. "Lunar Orbiter Photographs: Some Fundamental Observations," Science, Vol. 158 (December 22, 1967), pp. 1529-1535.

    The first three Lunar Orbiter spacecraft photographed 8% (600,000 square kilometers) of the near side of the Moon. High-resolution photographs show that the surface is dotted with a great number of small, perfectly circular craters from 50 meters diameter down to the limit of resolution. The majority of these are cup-shaped with distinctly sharp rims. But many also have shallow interiors and indistinct rims. The authors conclude that these craters were formed by primary and secondary impacts. Fresh craters are those which have material on the exterior slopes which is distinctly different from adjacent material of the inter-crater areas. These young craters also tend to have a profusion of angular blocks on the floors and exterior slopes. The albedo of these blocks and other ejecta material is relatively high. The number of fresh craters is much less than the number of craters not exhibiting these features.

Tyler, G. L., et al. "Bistatic-Radar Detection of Lunar Scattering Centers with Lunar Orbiter I," Science, Vol. 157 (July 14, 1967), pp. 193-195.

    Lunar Orbiter I bounced continuous-wave signals off of the Moon's surface, and these were received on Earth. Using the frequency spectrum and studying Doppler shifts, the investigators located discrete, heterogeneous scattering centers on the lunar surface. Shadowing, especially within five degrees of the terminator would effectively "hide" some scattering centers. On the other hand variations in surface reflectivity provide a model which will explain the observations. This could mean that material in scattering areas is considerably more compact or different from material in surrounding areas. The use of continuous-wave bistatic radar appears to offer a new method for mapping and study of lunar and planetary surfaces.

Ulrich, G. E. Advanced Systems Traverse Research Project Report with a Section on Problems for Geologic Investigations of the Orientale Region of the Moon by R. S. Saunders. United States Geological Survey Interagency Report: Astrogeology 7.

    This two-part report discusses some of the problems inherent in an extended lunar surface mission in the Orientale region and the scientific points of interest which such a mission might best help to explore. Lunar Orbiter photography played a significant role in the preparation of this report. The authors discuss various arguments about the origins of Orientale and the geological features which would be most significant in a surface investigation.

Urey, Harold C. "Mascons and the History of the Moon," Science, Vol. 162 (December 20, 1968), pp. 1408-1410.

    The Moon has a viscosity higher than that of Earth by a factor of 10^4. Mascons represent a non-isostatic condition in the surface of the Moon. Apparently an object collided with the Moon's surface, flattened out and left high-density material that has remained since the maria were formed. Lava flows cannot account for what is observed on the Moon. Maria areas on the Moon are not lava flows, and no liquid masses exist below the Moon. Thus large objects collided with the Moon in its early history. These objects should be similar to meteorites in composition and density. Finally, the Moon has sufficient rigidity to support these masses.

Urey, Harold C. "Water on the Moon," *Nature*, Vol. 216 (December 16, 1967), pp. 1094-1095.

    Urey summarizes several arguments against the presence of water on the Moon, and then he presents his own detailed argument, based upon his knowledge and new data from Lunar Orbiter photographs, in support of the presence of water on the Moon. The existence of rilles and of such landmarks as Schröter's Valley, the irregularities of the crater Krieger north of Aristarchus, and the knowledge of terrestrial geological processes causing pingos in areas of permafrost strongly support the theory that water has existed on the Moon and has caused various lunar surface formations. Urey defends the view that water, not lava or dust-gas mixtures, formed the maria and that these may yet be frozen seas. However, he concludes that this in no way defines the composition of the solid materials in the maria.

U.S. Army Topographic Command. *Final Report to National Aeronautics and Space Administration: Convergent Stereo Analysis*. Washington, D.C.: June 1969.

    This report, done under contract to NASA, explains the usefulness of stereoscopic photography transmitted to Earth by *Lunar Orbiters II, III*, and *V* in mapping the Moon. High-resolution stereo photographs include coverage otherwise unobtainable from a vertical mode. Moreover, the exaggerated height effects in convergent stereo photography should increase the accuracy in the determination of ground point elevations. The report discusses the problems of using existing computer programs and available photographic data for convergent photo triangulation. It also outlines the best methods for accomplishing triangulation. Tests with Lunar Orbiter data proved that accuracy of triangulation is increased by using high-resolution stereo photographs.

THE AUTHOR

While a graduate student at the University of Maryland, where he also taught courses in history, Bruce K. Byers devoted the summers of 1967 through 1970 to writing the Lunar Orbiter history as a summer intern at NASA. Earlier he had studied at the Ludwig-Maxmillians Universitaet in Munich, Germany. In 1971, Byers joined the U.S. Foreign Service, his first assignment taking him to Iran with the U.S. Information Service in 1972. His next assignment was as program officer with USIS in Bombay, India, where he now lives with his wife and three children. While in India, he has also written articles and lectured on the U.S. space program.

www.ingramcontent.com/pod-product-compliance
Lightning Source LLC
Chambersburg PA
CBHW081716170526
45167CB00009B/3594